An Introduction to Mechanical Engineering

An Introduction to Mechanical Engineering

Edited by
Brayden Anderson

Larsen & Keller
www.larsen-keller.com

An Introduction to Mechanical Engineering
Edited by Brayden Anderson
ISBN: 978-1-63549-179-1 (Hardback)

☰ Larsen & Keller

Published by Larsen and Keller Education,
5 Penn Plaza,
19th Floor,
New York, NY 10001, USA

Cataloging-in-Publication Data

An introduction to mechanical engineering / edited by Brayden Anderson.
   p. cm.
Includes bibliographical references and index.
ISBN 978-1-63549-179-1
1. Mechanical engineering. 2. Machinery.
3. Computer-aided engineering. I. Anderson, Brayden.
TJ145 .I58 2017
621--dc23

The publisher's policy is to use permanent paper from mills that operate a sustainable forestry policy. Furthermore, the publisher ensures that the text paper and cover boards used have met acceptable environmental accreditation standards.

Printed and bound in the United States of America.

For more information regarding Larsen and Keller Education and its products, please visit the publisher's website www.larsen-keller.com

Table of Contents

Permissions

Index

Preface

This book elucidates new techniques and applications of mechanical engineering in a multidisciplinary approach. It discusses in detail the various uses of this field in the present day scenario. Mechanical engineering refers to the practice of using the elements of material sciences, physics and engineering for the design, maintenance, construction and operation of mechanical systems. This textbook is compiled in such a manner, that it will provide in-depth knowledge about the theory and practice of mechanical engineering. Some of the diverse topics covered in it address the varied branches that fall under this category. This text is an essential guide for students in the fields of materials science, kinematics and industrial engineering.

A short introduction to every chapter is written below to provide an overview of the content of the book:

Chapter 1 - Mechanical engineering is a branch of engineering that is concerned with the principles of physics, analysis, manufacturing and maintenance of mechanical systems. It involves the design and production of machines. The chapter on mechanical engineering offers an insightful focus, keeping in mind the complex subject matter; **Chapter 2** - Mechanics, kinematics and structural analysis are some significant topics related to mechanical engineering. Mechanics is an area of science that deals with the behavior of physical bodies. The following chapter unfolds the main concepts and aspects of mechanical engineering in a critical manner; **Chapter 3 -** This chapter elucidates the principles of mechanical engineering, providing a complete understanding on the subject. The principles of mechanical engineering elucidated in this chapter are computer-aided engineering, solid modeling, computer-aided design and computational fluid dynamics. The section strategically encompasses all the principles of mechanical engineering, providing a complete understanding; **Chapter 4** - Mechanical engineering uses various design models upon which new creations are based. Some of the basic models and ideas used in this field are kinematic chain, rigid body dynamics, equations of motion etc. This chapter explains the fundamentals of mechanical engineering in an easy and comprehensive manner; **Chapter 5** - Mechatronics is a field of science that includes subjects such as computer engineering, telecommunications engineering and control engineering along with mechanical engineering. The various sub disciplines explained in this section are mechatronics, HVAC, robotics and tribology. This chapter will provide a glimpse of the sub disciplines of mechanical engineering; **Chapter 6** - The significant aspects of mechanism engineering are discussed in this section. Mechanism is a device designed to transform input forces and movement into a desired set of output forces and movement. The aspects elucidated in this chapter are of vital importance, and provide a better understanding of mechanism engineering.

Finally, I would like to thank my fellow scholars who gave constructive feedback and my family members who supported me at every step.

Editor

Introduction to Mechanical Engineering

Mechanical engineering is a branch of engineering that is concerned with the principles of physics, analysis, manufacturing and maintenance of mechanical systems. It involves the design and production of machines. The chapter on mechanical engineering offers an insightful focus, keeping in mind the complex subject matter.

Mechanical Engineering

Mechanical engineering is the discipline that applies the principles of engineering, physics, and materials science for the design, analysis, manufacturing, and maintenance of mechanical systems. It is the branch of engineering that involves the design, production, and operation of machinery. It is one of the oldest and broadest of the engineering disciplines.

The engineering field requires an understanding of core concepts including mechanics, kinematics, thermodynamics, materials science, structural analysis, and electricity. Mechanical engineers use these core principles along with tools like computer-aided design, and product lifecycle management to design and analyze manufacturing plants, industrial equipment and machinery, heating and cooling systems, transport systems, aircraft, watercraft, robotics, medical devices, weapons, and others.

W16 engine of the Bugatti Veyron. Mechanical engineers design and build engines, power plants, other machines...

Mechanical engineering emerged as a field during the Industrial Revolution in Europe in the 18th century; however, its development can be traced back several thousand years around the world. Mechanical engineering science emerged in the 19th century as a result of developments in the field of physics. The field has continually evolved to incorporate advancements in technology, and mechanical engineers today are pursuing developments in such fields as composites, mechatronics, and nanotechnology. Mechanical engineering overlaps with aerospace engineering, metallurgical engineering, civil engineering, electrical engineering, manufacturing engineering, chemical engineering, industrial engineering, and other engineering disciplines to varying amounts. Mechanical engineers may also work in the field of biomedical engineering, specifically with biomechanics, transport phenomena, biomechatronics, bionanotechnology, and modeling of biological systems.

...structures, and vehicles of all sizes.

History

Mechanical engineering finds its application in the archives of various ancient and medieval societies throughout mankind. In ancient Greece, the works of Archimedes (287–212 BC) deeply influenced mechanics in the Western tradition and Heron of Alexandria (c. 10–70 AD) created the first steam engine (Aeolipile). In China, Zhang Heng (78–139 AD) improved a water clock and invented a seismometer, and Ma Jun (200–265 AD) invented a chariot with differential gears. The medieval Chinese horologist and engineer Su Song (1020–1101 AD) incorporated an escapement mechanism into his astronomical clock tower two centuries before any escapement can be found in clocks of medieval Europe, as well as the world's first known endless power-transmitting chain drive.

During the years from 7th to 15th century, the era called the Islamic Golden Age, there were remarkable contributions from Muslim inventors in the field of mechanical technology. Al-Jazari, who was one of them, wrote his famous *Book of Knowledge of Ingenious Mechanical Devices* in 1206, and presented many mechanical designs. He is also considered to be the inventor of such mechanical devices which now form the very basic of mechanisms, such as the crankshaft and camshaft.

Important breakthroughs in the foundations of mechanical engineering occurred in England during the 17th century when Sir Isaac Newton both formulated the three Newton's Laws of Motion and developed Calculus, the mathematical basis of physics. Newton was reluctant to publish his methods and laws for years, but he was finally persuaded to do so by his colleagues, such as Sir Edmund Halley, much to the benefit of all mankind. Gottfried Wilhelm Leibniz is also credited with creating Calculus during the same time frame.

During the early 19th century in England, Germany and Scotland, the development of machine tools led mechanical engineering to develop as a separate field within engineering, providing manufacturing machines and the engines to power them. The first British professional society of mechanical engineers was formed in 1847 Institution of Mechanical Engineers, thirty years after the civil engineers formed the first such professional society Institution of Civil Engineers. On the European continent, Johann von Zimmermann (1820–1901) founded the first factory for grinding machines in Chemnitz, Germany in 1848.

In the United States, the American Society of Mechanical Engineers (ASME) was formed in 1880, becoming the third such professional engineering society, after the American Society of Civil Engineers (1852) and the American Institute of Mining Engineers (1871). The first schools in the United States to offer an engineering education were the United States Military Academy in 1817, an institution now known as Norwich University in 1819, and Rensselaer Polytechnic Institute in 1825. Education in mechanical engineering has historically been based on a strong foundation in mathematics and science.

Education

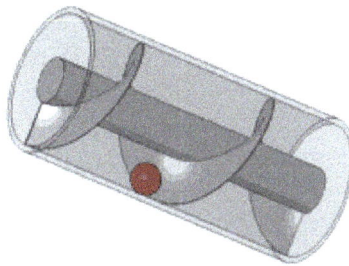

Archimedes' screw was operated by hand and could efficiently raise water, as the animated red ball demonstrates.

Degrees in mechanical engineering are offered at various universities worldwide. In Brazil, Ireland, Philippines, Pakistan, China, Greece, Turkey, North America, South Asia, Nepal, India, Dominican Republic and the United Kingdom, mechanical engineering programs typically take four to five years of study and result in a Bachelor of Engineering (B.Eng. or B.E.), Bachelor of Science (B.Sc. or B.S.), Bachelor of Science Engineering (B.Sc.Eng.), Bachelor of Technology (B.Tech.), Bachelor of Mechanical Engineering (B.M.E.), or Bachelor of Applied Science (B.A.Sc.) degree, in or with emphasis in mechanical engineering. In Spain, Portugal and most of South America, where neither B.Sc. nor B.Tech. programs have been adopted, the formal name for the degree is "Mechanical Engineer", and the course work is based on five or six years of training. In Italy the course work is based on five years of training, but in order to qualify as an Engineer one has to pass a state exam at the end of the course. In Greece, the coursework is based on a five-year curriculum and the requirement of a 'Diploma' Thesis, which upon completion a 'Diploma' is awarded rather than a B.Sc.

In Australia, mechanical engineering degrees are awarded as Bachelor of Engineering (Mechanical) or similar nomenclature although there are an increasing number of specialisations. The degree takes four years of full-time study to achieve. To ensure quality in engineering degrees, Engineers Australia accredits engineering degrees awarded by Australian universities in accordance with the global Washington Accord. Before the degree can be awarded, the student must complete at least 3 months of on the job work experience in an engineering firm. Similar systems are also present in South Africa and are overseen by the Engineering Council of South Africa (ECSA).

In the United States, most undergraduate mechanical engineering programs are accredited by the

Accreditation Board for Engineering and Technology (ABET) to ensure similar course requirements and standards among universities. The ABET web site lists 302 accredited mechanical engineering programs as of 11 March 2014. Mechanical engineering programs in Canada are accredited by the Canadian Engineering Accreditation Board (CEAB), and most other countries offering engineering degrees have similar accreditation societies.

In India, to become an engineer, one need to have an engineering degree like a B.Tech or B.E or have a diploma in engineering or by completing a course in an engineering trade like fitter from the Industrial Training Institute (ITIs) to receive a "ITI Trade Certificate" and also have to pass the All India Trade Test (AITT) with an engineering trade conducted by the National Council of Vocational Training (NCVT) by which one is awarded a "National Trade Certificate". Similar systems are used in Nepal.

Some mechanical engineers go on to pursue a postgraduate degree such as a Master of Engineering, Master of Technology, Master of Science, Master of Engineering Management (M.Eng.Mgt. or M.E.M.), a Doctor of Philosophy in engineering (Eng.D. or Ph.D.) or an engineer's degree. The master's and engineer's degrees may or may not include research. The Doctor of Philosophy includes a significant research component and is often viewed as the entry point to academia. The Engineer's degree exists at a few institutions at an intermediate level between the master's degree and the doctorate.

Coursework

Standards set by each country's accreditation society are intended to provide uniformity in fundamental subject material, promote competence among graduating engineers, and to maintain confidence in the engineering profession as a whole. Engineering programs in the U.S., for example, are required by ABET to show that their students can "work professionally in both thermal and mechanical systems areas." The specific courses required to graduate, however, may differ from program to program. Universities and Institutes of technology will often combine multiple subjects into a single class or split a subject into multiple classes, depending on the faculty available and the university's major area(s) of research.

The fundamental subjects of mechanical engineering usually include:

- Mathematics (in particular, calculus, differential equations, and linear algebra)
- Basic physical sciences (including physics and chemistry)
- Statics and dynamics
- Strength of materials and solid mechanics
- Materials Engineering, Composites
- Thermodynamics, heat transfer, energy conversion, and HVAC
- Fuels, combustion, Internal combustion engine
- Fluid mechanics (including fluid statics and fluid dynamics)
- Mechanism and Machine design (including kinematics and dynamics)

- Instrumentation and measurement

- Manufacturing engineering, technology, or processes

- Vibration, control theory and control engineering

- Hydraulics, and pneumatics

- Mechatronics, and robotics

- Engineering design and product design

- Drafting, computer-aided design (CAD) and computer-aided manufacturing (CAM)

Mechanical engineers are also expected to understand and be able to apply basic concepts from chemistry, physics, chemical engineering, civil engineering, and electrical engineering. All mechanical engineering programs include multiple semesters of mathematical classes including calculus, and advanced mathematical concepts including differential equations, partial differential equations, linear algebra, abstract algebra, and differential geometry, among others.

In addition to the core mechanical engineering curriculum, many mechanical engineering programs offer more specialized programs and classes, such as control systems, robotics, transport and logistics, cryogenics, fuel technology, automotive engineering, biomechanics, vibration, optics and others, if a separate department does not exist for these subjects.

Most mechanical engineering programs also require varying amounts of research or community projects to gain practical problem-solving experience. In the United States it is common for mechanical engineering students to complete one or more internships while studying, though this is not typically mandated by the university. Cooperative education is another option. Future work skills research puts demand on study components that feed student's creativity and innovation.

License and Regulation

Engineers may seek license by a state, provincial, or national government. The purpose of this process is to ensure that engineers possess the necessary technical knowledge, real-world experience, and knowledge of the local legal system to practice engineering at a professional level. Once certified, the engineer is given the title of Professional Engineer (in the United States, Canada, Japan, South Korea, Bangladesh and South Africa), Chartered Engineer (in the United Kingdom, Ireland, India and Zimbabwe), *Chartered Professional Engineer* (in Australia and New Zealand) or *European Engineer* (much of the European Union), Registered Engineer or Professional Engineer in Philippines and Pakistan.

In the U.S., to become a licensed Professional Engineer (PE), an engineer must pass the comprehensive FE (Fundamentals of Engineering) exam, work a minimum of 4 years as an *Engineering Intern (EI)* or *Engineer-in-Training (EIT)*, and pass the "Principles and Practice" or PE (Practicing Engineer or Professional Engineer) exams. The requirements and steps of this process are set forth by the National Council of Examiners for Engineering and Surveying (NCEES), a composed of engineering and land surveying licensing boards representing all U.S. states and territories.

In the UK, current graduates require a BEng plus an appropriate master's degree or an integrated MEng degree, a minimum of 4 years post graduate on the job competency development, and a peer reviewed project report in the candidates specialty area in order to become a Chartered Mechanical Engineer (CEng, MIMechE) through the Institution of Mechanical Engineers. CEng MIMechE can also be obtained via an examination route.

In most developed countries, certain engineering tasks, such as the design of bridges, electric power plants, and chemical plants, must be approved by a professional engineer or a chartered engineer. "Only a licensed engineer, for instance, may prepare, sign, seal and submit engineering plans and drawings to a public authority for approval, or to seal engineering work for public and private clients." This requirement can be written into state and provincial legislation, such as in the Canadian provinces, for example the Ontario or Quebec's Engineer Act.

In other countries, such as Australia, and the UK, no such legislation exists; however, practically all certifying bodies maintain a code of ethics independent of legislation, that they expect all members to abide by or risk expulsion.

Salaries and Workforce Statistics

The total number of engineers employed in the U.S. in 2009 was roughly 1.6 million. Of these, 239,000 were mechanical engineers (14.9%), the second largest discipline by size behind civil (278,000). The total number of mechanical engineering jobs in 2009 was projected to grow 6% over the next decade, with average starting salaries being $58,800 with a bachelor's degree. The median annual income of mechanical engineers in the U.S. workforce was $80,580. The median income was highest when working for the government ($92,030), and lowest in education ($57,090) as of 2012.

Modern Tools

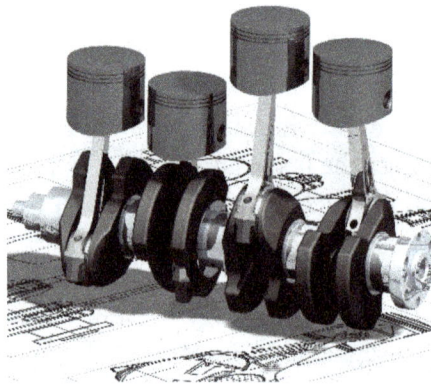

An oblique view of a four-cylinder inline crankshaft with pistons

Many mechanical engineering companies, especially those in industrialized nations, have begun to incorporate computer-aided engineering (CAE) programs into their existing design and analysis processes, including 2D and 3D solid modeling computer-aided design (CAD). This method has many benefits, including easier and more exhaustive visualization of products, the ability to create virtual assemblies of parts, and the ease of use in designing mating interfaces and tolerances.

Other CAE programs commonly used by mechanical engineers include product lifecycle management (PLM) tools and analysis tools used to perform complex simulations. Analysis tools may be used to predict product response to expected loads, including fatigue life and manufacturability. These tools include finite element analysis (FEA), computational fluid dynamics (CFD), and computer-aided manufacturing (CAM).

Using CAE programs, a mechanical design team can quickly and cheaply iterate the design process to develop a product that better meets cost, performance, and other constraints. No physical prototype need be created until the design nears completion, allowing hundreds or thousands of designs to be evaluated, instead of a relative few. In addition, CAE analysis programs can model complicated physical phenomena which cannot be solved by hand, such as viscoelasticity, complex contact between mating parts, or non-Newtonian flows.

As mechanical engineering begins to merge with other disciplines, as seen in mechatronics, multidisciplinary design optimization (MDO) is being used with other CAE programs to automate and improve the iterative design process. MDO tools wrap around existing CAE processes, allowing product evaluation to continue even after the analyst goes home for the day. They also utilize sophisticated optimization algorithms to more intelligently explore possible designs, often finding better, innovative solutions to difficult multidisciplinary design problems.

Subdisciplines

The field of mechanical engineering can be thought of as a collection of many mechanical engineering science disciplines. Several of these subdisciplines which are typically taught at the undergraduate level are listed below, with a brief explanation and the most common application of each. Some of these subdisciplines are unique to mechanical engineering, while others are a combination of mechanical engineering and one or more other disciplines. Most work that a mechanical engineer does uses skills and techniques from several of these subdisciplines, as well as specialized subdisciplines. Specialized subdisciplines, as used in this article, are more likely to be the subject of graduate studies or on-the-job training than undergraduate research. Several specialized subdisciplines are discussed in this section.

Mechanics

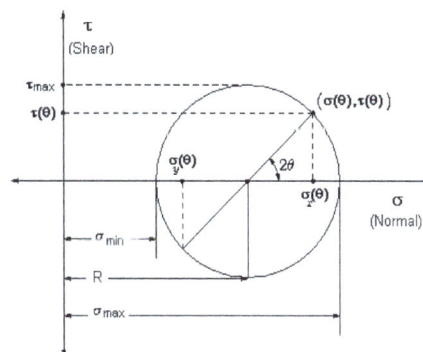

Mohr's circle, a common tool to study stresses in a mechanical element

Mechanics is, in the most general sense, the study of forces and their effect upon matter. Typically, engineering mechanics is used to analyze and predict the acceleration and deformation (both elastic and plastic) of objects under known forces (also called loads) or stresses. Subdisciplines of mechanics include

- Statics, the study of non-moving bodies under known loads, how forces affect static bodies

- Dynamics (or kinetics), the study of how forces affect moving bodies

- Mechanics of materials, the study of how different materials deform under various types of stress

- Fluid mechanics, the study of how fluids react to forces

- Kinematics, the study of the motion of bodies (objects) and systems (groups of objects), while ignoring the forces that cause the motion. Kinematics is often used in the design and analysis of mechanisms.

- Continuum mechanics, a method of applying mechanics that assumes that objects are continuous (rather than discrete)

Mechanical engineers typically use mechanics in the design or analysis phases of engineering. If the engineering project were the design of a vehicle, statics might be employed to design the frame of the vehicle, in order to evaluate where the stresses will be most intense. Dynamics might be used when designing the car's engine, to evaluate the forces in the pistons and cams as the engine cycles. Mechanics of materials might be used to choose appropriate materials for the frame and engine. Fluid mechanics might be used to design a ventilation system for the vehicle, or to design the intake system for the engine.

Mechatronics and Robotics

Training FMS with learning robot SCORBOT-ER 4u, workbench CNC Mill and CNC Lathe

Mechatronics is a combination of mechanics and electronics. It is an interdisciplinary branch of mechanical engineering, electrical engineering and software engineering that is concerned with integrating electrical and mechanical engineering to create hybrid systems. In this way, machines can be automated through the use of electric motors, servo-mechanisms, and other electrical systems in conjunction with special software. A common example of a mechatronics system is a CD-ROM drive. Mechanical systems open and close the drive, spin the CD and move the laser, while an optical system reads the data on the CD and converts it to bits. Integrated software controls the process and communicates the contents of the CD to the computer.

Robotics is the application of mechatronics to create robots, which are often used in industry to perform tasks that are dangerous, unpleasant, or repetitive. These robots may be of any shape and size, but all are preprogrammed and interact physically with the world. To create a robot, an engineer typically employs kinematics (to determine the robot's range of motion) and mechanics (to determine the stresses within the robot).

Robots are used extensively in industrial engineering. They allow businesses to save money on labor, perform tasks that are either too dangerous or too precise for humans to perform them economically, and to ensure better quality. Many companies employ assembly lines of robots, especially in Automotive Industries and some factories are so robotized that they can run by themselves. Outside the factory, robots have been employed in bomb disposal, space exploration, and many other fields. Robots are also sold for various residential applications, from recreation to domestic applications.

Structural Analysis

Structural analysis is the branch of mechanical engineering (and also civil engineering) devoted to examining why and how objects fail and to fix the objects and their performance. Structural failures occur in two general modes: static failure, and fatigue failure. *Static structural failure* occurs when, upon being loaded (having a force applied) the object being analyzed either breaks or is deformed plastically, depending on the criterion for failure. *Fatigue failure* occurs when an object fails after a number of repeated loading and unloading cycles. Fatigue failure occurs because of imperfections in the object: a microscopic crack on the surface of the object, for instance, will grow slightly with each cycle (propagation) until the crack is large enough to cause ultimate failure.

Failure is not simply defined as when a part breaks, however; it is defined as when a part does not operate as intended. Some systems, such as the perforated top sections of some plastic bags, are designed to break. If these systems do not break, failure analysis might be employed to determine the cause.

Structural analysis is often used by mechanical engineers after a failure has occurred, or when designing to prevent failure. Engineers often use online documents and books such as those published by ASM to aid them in determining the type of failure and possible causes.

Structural analysis may be used in the office when designing parts, in the field to analyze failed parts, or in laboratories where parts might undergo controlled failure tests.

Thermodynamics and Thermo-Science

Thermodynamics is an applied science used in several branches of engineering, including mechanical and chemical engineering. At its simplest, thermodynamics is the study of energy, its use and transformation through a system. Typically, engineering thermodynamics is concerned with changing energy from one form to another. As an example, automotive engines convert chemical energy (enthalpy) from the fuel into heat, and then into mechanical work that eventually turns the wheels.

Thermodynamics principles are used by mechanical engineers in the fields of heat transfer, thermofluids, and energy conversion. Mechanical engineers use thermo-science to design engines and power plants, heating, ventilation, and air-conditioning (HVAC) systems, heat exchangers, heat sinks, radiators, refrigeration, insulation, and others.

Design and Drafting

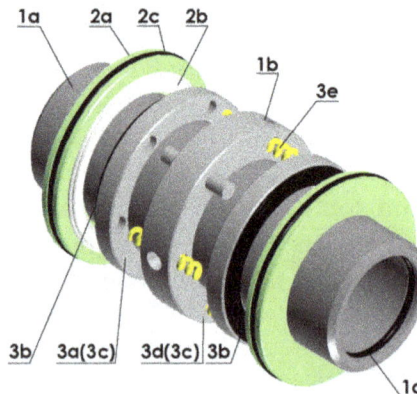

A CAD model of a mechanical double seal

Drafting or technical drawing is the means by which mechanical engineers design products and create instructions for manufacturing parts. A technical drawing can be a computer model or hand-drawn schematic showing all the dimensions necessary to manufacture a part, as well as assembly notes, a list of required materials, and other pertinent information. A U.S. mechanical engineer or skilled worker who creates technical drawings may be referred to as a drafter or draftsman. Drafting has historically been a two-dimensional process, but computer-aided design (CAD) programs now allow the designer to create in three dimensions.

Instructions for manufacturing a part must be fed to the necessary machinery, either manually, through programmed instructions, or through the use of a computer-aided manufacturing (CAM) or combined CAD/CAM program. Optionally, an engineer may also manually manufacture a part using the technical drawings, but this is becoming an increasing rarity, with the advent of computer numerically controlled (CNC) manufacturing. Engineers primarily manually manufacture parts in the areas of applied spray coatings, finishes, and other processes that cannot economically or practically be done by a machine.

Drafting is used in nearly every subdiscipline of mechanical engineering, and by many other branches of engineering and architecture. Three-dimensional models created using CAD software are also commonly used in finite element analysis (FEA) and computational fluid dynamics (CFD).

Areas of Research

Mechanical engineers are constantly pushing the boundaries of what is physically possible in order to produce safer, cheaper, and more efficient machines and mechanical systems. Some technologies at the cutting edge of mechanical engineering are listed below.

Micro Electro-Mechanical Systems (MEMS)

Micron-scale mechanical components such as springs, gears, fluidic and heat transfer devices are fabricated from a variety of substrate materials such as silicon, glass and polymers like SU8. Examples of MEMS components are the accelerometers that are used as car airbag sensors, modern cell phones, gyroscopes for precise positioning and microfluidic devices used in biomedical applications.

Friction Stir Welding (FSW)

Friction stir welding, a new type of welding, was discovered in 1991 by The Welding Institute (TWI). The innovative steady state (non-fusion) welding technique joins materials previously un-weldable, including several aluminum alloys. It plays an important role in the future construction of airplanes, potentially replacing rivets. Current uses of this technology to date include welding the seams of the aluminum main Space Shuttle external tank, Orion Crew Vehicle test article, Boeing Delta II and Delta IV Expendable Launch Vehicles and the SpaceX Falcon 1 rocket, armor plating for amphibious assault ships, and welding the wings and fuselage panels of the new Eclipse 500 aircraft from Eclipse Aviation among an increasingly growing pool of uses.

Composites

Composite cloth consisting of woven carbon fiber

Composites or composite materials are a combination of materials which provide different physical characteristics than either material separately. Composite material research within mechanical engineering typically focuses on designing (and, subsequently, finding applications for) stronger or more rigid materials while attempting to reduce weight, susceptibility to corrosion, and other undesirable factors. Carbon fiber reinforced composites, for instance, have been used in such diverse applications as spacecraft and fishing rods.

Mechatronics

Mechatronics is the synergistic combination of mechanical engineering, electronic engineering, and software engineering. The purpose of this interdisciplinary engineering field is the study of automation from an engineering perspective and serves the purposes of controlling advanced hybrid systems.

Nanotechnology

At the smallest scales, mechanical engineering becomes nanotechnology—one speculative goal of which is to create a molecular assembler to build molecules and materials via mechanosynthesis. For now that goal remains within exploratory engineering. Areas of current mechanical engineering research in nanotechnology include nanofilters, nanofilms, and nanostructures, among others.

Finite Element Analysis

This field is not new, as the basis of Finite Element Analysis (FEA) or Finite Element Method (FEM) dates back to 1941. But evolution of computers has made FEA/FEM a viable option for analysis of structural problems. Many commercial codes such as ANSYS, NASTRAN and ABAQUS are widely used in industry for research and the design of components. Some 3D modeling and CAD software packages have added FEA modules.

Other techniques such as finite difference method (FDM) and finite-volume method (FVM) are employed to solve problems relating heat and mass transfer, fluid flows, fluid surface interaction etc.

Biomechanics

Biomechanics is the application of mechanical principles to biological systems, such as humans, animals, plants, organs, and cells. Biomechanics also aids in creating prosthetic limbs and artificial organs for humans.

Biomechanics is closely related to engineering, because it often uses traditional engineering sciences to analyse biological systems. Some simple applications of Newtonian mechanics and/or materials sciences can supply correct approximations to the mechanics of many biological systems.

Over the past decade the Finite element method (FEM) has also entered the Biomedical sector highlighting further engineering aspects of Biomechanics. FEM has since then established itself as an alternative to in vivo surgical assessment and gained the wide acceptance of academia. The main advantage of Computational Biomechanics lies in its ability to determine the endo-anatomical response of an anatomy, without being subject to ethical restrictions. This has led FE modelling to the point of becoming ubiquitous in several fields of Biomechanics while several projects have even adopted an open source philosophy (e.g. BioSpine).

Computational Fluid Dynamics

Computational fluid dynamics, usually abbreviated as CFD, is a branch of fluid mechanics that uses numerical methods and algorithms to solve and analyze problems that involve fluid flows. Computers are used to perform the calculations required to simulate the interaction of liquids and gases with surfaces defined by boundary conditions. With high-speed supercomputers, better solutions can be achieved. Ongoing research yields software that improves the accuracy and speed of complex simulation scenarios such as transonic or turbulent flows. Initial validation of such software is performed using a wind tunnel with the final validation coming in full-scale testing, e.g. flight tests.

Acoustical Engineering

Acoustical engineering is one of many other sub disciplines of mechanical engineering and is the application of acoustics. Acoustical engineering is the study of Sound and Vibration. These engineers work effectively to reduce noise pollution in mechanical devices and in buildings by soundproofing or removing sources of unwanted noise. The study of acoustics can range from designing a more efficient hearing aid, microphone, headphone, or recording studio to enhancing the sound quality of an orchestra hall. Acoustical engineering also deals with the vibration of different mechanical systems.

Related Fields

Manufacturing engineering, Aerospace engineering and Automotive engineering are sometimes grouped with mechanical engineering. A bachelor's degree in these areas will typically have a difference of a few specialized classes.

Machine

A machine is a tool containing one or more parts that uses energy to perform an intended action. Machines are usually powered by mechanical, chemical, thermal, or electrical means, and are often motorized. Historically, a power tool also required moving parts to classify as a machine. However, the advent of electronics has led to the development of power tools without moving parts that are considered machines.

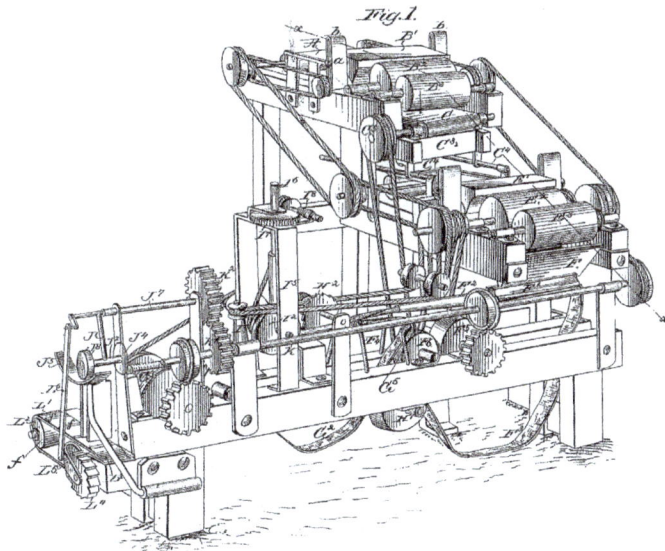

James Albert Bonsack's cigarette rolling machine, invented in 1880 and patented in 1881.

A simple machine is a device that simply transforms the direction or magnitude of a force, but a large number of more complex machines exist. Examples include vehicles, electronic systems, molecular machines, computers, television, and radio.

Etymology

This meaning is found in late medieval French, and is adopted from the French into English in the mid-16th century.

In the 17th century, the word could also mean a scheme or plot, a meaning now expressed by the derived machination. The modern meaning develops out of specialized application of the term to stage engines used in theater and to military siege engines, both in the late 16th and early 17th centuries. The OED traces the formal, modern meaning to John Harris' *Lexicon Technicum* (1704), which has:

Machine, or Engine, in Mechanicks, is whatsoever hath Force sufficient either to raise or stop the Motion of a Body... Simple Machines are commonly reckoned to be Six in Number, viz. the Ballance, Leaver, Pulley, Wheel, Wedge, and Screw... Compound Machines, or Engines, are innumerable.

The word *engine* used as a (near-)synonym both by Harris and in later language derives ultimately (via Old French) from Latin *ingenium* "ingenuity, an invention".

History

Flint hand axe found in Winchester

Perhaps the first example of a human made device designed to manage power is the hand axe, made by chipping flint to form a wedge. A wedge is a simple machine that transforms lateral force and movement of the tool into a transverse splitting force and movement of the workpiece.

The idea of a *simple machine* originated with the Greek philosopher Archimedes around the 3rd century BC, who studied the Archimedean simple machines: lever, pulley, and screw. He discovered the principle of mechanical advantage in the lever. Later Greek philosophers defined the classic five simple machines (excluding the inclined plane) and were able to roughly calculate their mechanical advantage. Heron of Alexandria (ca. 10–75 AD) in his work *Mechanics* lists five mechanisms that can "set a load in motion"; lever, windlass, pulley, wedge, and screw, and describes their fabrication and uses. However the Greeks' understanding was limited to statics (the balance of forces) and did not include dynamics (the tradeoff between force and distance) or the concept of work.

During the Renaissance the dynamics of the *Mechanical Powers*, as the simple machines were called, began to be studied from the standpoint of how much useful work they could perform, leading eventually to the new concept of mechanical work. In 1586 Flemish engineer Simon Stevin derived the mechanical advantage of the inclined plane, and it was included with the other simple machines. The complete dynamic theory of simple machines was worked out by Italian scientist Galileo Galilei in 1600 in *Le Meccaniche* ("On Mechanics"). He was the first to understand that simple machines do not create energy, they merely transform it.

The classic rules of sliding friction in machines were discovered by Leonardo da Vinci (1452–1519), but remained unpublished in his notebooks. They were rediscovered by Guillaume Amontons (1699) and were further developed by Charles-Augustin de Coulomb (1785).

Types

Types of machines and related components	
Classification	**Machine(s)**
Simple machines	Inclined plane, Wheel and axle, Lever, Pulley, Wedge, Screw
Mechanical components	Axle, Bearings, Belts, Bucket, Fastener, Gear, Key, Link chains, Rack and pinion, Roller chains, Rope, Seals, Spring, Wheel
Clock	Atomic clock, Watch, Pendulum clock, Quartz clock
Compressors and Pumps	Archimedes' screw, Eductor-jet pump, Hydraulic ram, Pump, Trompe, Vacuum pump
Heat engines — External combustion engines	Steam engine, Stirling engine
Heat engines — Internal combustion engines	Reciprocating engine, Gas turbine
Heat pumps	Absorption refrigerator, Thermoelectric refrigerator, Regenerative cooling
Linkages	Pantograph, Cam, Peaucellier-Lipkin
Turbine	Gas turbine, Jet engine, Steam turbine, Water turbine, Wind generator, Windmill
Aerofoil	Sail, Wing, Rudder, Flap, Propeller
Information technology	Computer, Calculator, Telecommunications networks
Electricity	Vacuum tube, Transistor, Diode, Resistor, Capacitor, Inductor, Memristor, Semiconductor
Robots	Actuator, Servo, Servomechanism, Stepper motor
Miscellaneous	Vending machine, Wind tunnel, Check weighing machines, Riveting machines

Mechanical

The word mechanical refers to the work that has been produced by machines or the machinery. It mostly relates to the machinery tools and the mechanical applications of science. Some of its synonyms are automatic and mechanic.

Simple Machines

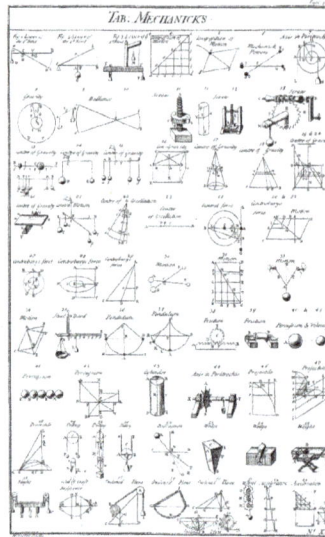

Table of simple mechanisms, from *Chambers' Cyclopædia*, 1728. Simple machines provide a "vocabulary" for understanding more complex machines.

The idea that a machine can be broken down into simple movable elements led Archimedes to define the lever, pulley and screw as simple machines. By the time of the Renaissance this list increased to include the wheel and axle, wedge and inclined plane.

Engines

An engine or motor is a machine designed to convert energy into useful mechanical motion. Heat engines, including internal combustion engines and external combustion engines (such as steam engines) burn a fuel to create heat, which is then used to create motion. Electric motors convert electrical energy into mechanical motion, pneumatic motors use compressed air and others, such as wind-up toys use elastic energy. In biological systems, molecular motors like myosins in muscles use chemical energy to create motion.

Electrical

Electrical means operating by or producing electricity, relating to or concerned with electricity. In other words, it means using, providing, producing, transmitting or operated by electricity.

Electrical Machine

An electrical machine is the generic name for a device that converts mechanical energy to electrical energy, converts electrical energy to mechanical energy, or changes alternating current from one voltage level to a different voltage level.

Electronic Machine

Electronics is the branch of physics, engineering and technology dealing with electrical circuits that involve active electrical components such as vacuum tubes, transistors, diodes and integrated circuits, and associated passive interconnection technologies. The nonlinear behaviour of active components and their ability to control electron flows makes amplification of weak signals possible and is usually applied to information and signal processing. Similarly, the ability of electronic devices to act as switches makes digital information processing possible. Interconnection technologies such as circuit boards, electronic packaging technology, and other varied forms of communication infrastructure complete circuit functionality and transform the mixed components into a working system.

Computing Machines

Computers are machines to process information, often in the form of numbers. Charles Babbage designed various machines to tabulate logarithms and other functions in 1837. His Difference engine can be considered an advanced mechanical calculator and his Analytical Engine a forerunner of the modern computer, though none were built in Babbage's lifetime.

Modern computers are electronic ones. They use electric charge, current or magnetization to store and manipulate information. Computer architecture deals with detailed design of computers. There are also simplified models of computers, like State machine and Turing machine.

Molecular Machines

Study of the molecules and proteins that are the basis of biological functions has led to the concept of a molecular machine. For example, current models of the operation of the kinesin molecule that transports vesicles inside the cell as well as the myosin molecule that operates against actin to cause muscle contraction; these molecules control movement in response to chemical stimuli.

Researchers in nano-technology are working to construct molecules that perform movement in response to a specific stimulus. In contrast to molecules such as kinesin and myosin, these nano-machines or molecular machines are constructions like traditional machines that are designed to perform in a task.

Machine Elements

Machines are assembled from standardized types of components. These elements consist of mechanisms that control movement in various ways such as gear trains, transistor switches, belt or chain drives, linkages, cam and follower systems, brakes and clutches, and *structural components* such as frame members and fasteners.

Modern machines include sensors, actuators and computer controllers. The shape, texture and color of covers provide a styling and operational interface between the mechanical components of a machine and its users.

Mechanisms

Assemblies within a machine that control movement are often called "mechanisms." Mechanisms are generally classified as gears and gear trains, cam and follower mechanisms, and linkages, though there are other special mechanisms such as clamping linkages, indexing mechanisms and friction devices such as brakes and clutches.

Controllers

Controllers combine sensors, logic, and actuators to maintain the performance of components of a machine. Perhaps the best known is the flyball governor for a steam engine. Examples of these devices range from a thermostat that as temperature rises opens a valve to cooling water to speed controllers such the cruise control system in an automobile. The programmable logic controller replaced relays and specialized control mechanisms with a programmable computer. Servo motors that accurately position a shaft in response to an electrical command are the actuators that make robotic systems possible.

Impact

Industrial Revolution

The Industrial Revolution was a period from 1750 to 1850 where changes in agriculture, manufacturing, mining, transportation, and technology had a profound effect on the social, economic and cultural conditions of the times. It began in the United Kingdom, then subsequently spread throughout Western Europe, North America, Japan, and eventually the rest of the world.

Starting in the later part of the 18th century, there began a transition in parts of Great Britain's previously manual labour and draft-animal–based economy towards machine-based manufacturing. It started with the mechanisation of the textile industries, the development of iron-making techniques and the increased use of refined coal.

Mechanization and Automation

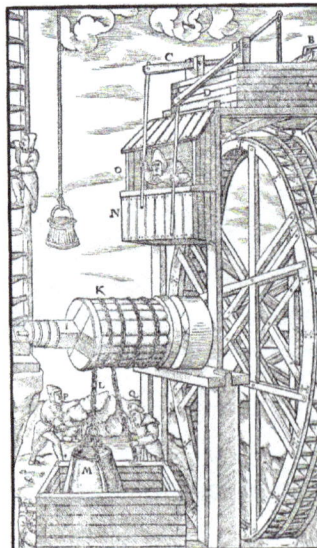

A water-powered mine hoist used for raising ore. This woodblock is from De re metallica by Georg Bauer (Latinized name Georgius Agricola, ca. 1555) an early mining textbook that contains numerous drawings and descriptions of mining equipment.

Mechanization or mechanisation (BE) is providing human operators with machinery that assists them with the muscular requirements of work or displaces muscular work. In some fields, mechanization includes the use of hand tools. In modern usage, such as in engineering or economics, mechanization implies machinery more complex than hand tools and would not include simple devices such as an un-geared horse or donkey mill. Devices that cause speed changes or changes to or from reciprocating to rotary motion, using means such as gears, pulleys or sheaves and belts, shafts, cams and cranks, usually are considered machines. After electrification, when most small machinery was no longer hand powered, mechanization was synonymous with motorized machines.

Automation is the use of control systems and information technologies to reduce the need for human work in the production of goods and services. In the scope of industrialization, automation is a step beyond mechanization. Whereas mechanization provides human operators with machinery to assist them with the muscular requirements of work, automation greatly decreases the need for human sensory and mental requirements as well. Automation plays an increasingly important role in the world economy and in daily experience.

The *Digesting Duck* by Jacques de Vaucanson, hailed in 1739 as the first automaton capable of digestion

Automata

An automaton (plural: automata or automatons) is a self-operating machine. The word is sometimes used to describe a robot, more specifically an autonomous robot. An alternative spelling, now obsolete, is automation.

Machine (Mechanical)

Diesel engine, friction clutch and gear transmission of an automobile.

Machines employ power to achieve desired forces and movement (motion). A machine has a power source and actuators that generate forces and movement, and a system of mechanisms that shape the actuator input to achieve a specific application of output forces and movement. Modern machines often include computers and sensors that monitor performance and plan movement, and are called mechanical systems.

The meaning of the word "machine" is traced by the Oxford English Dictionary to an independently functioning structure and by Merriam-Webster Dictionary to something that has been constructed. This includes human design into the meaning of machine.

The adjective "mechanical" refers to skill in the practical application of an art or science, as well as relating to or caused by movement, physical forces, properties or agents such as is dealt with by mechanics. Similarly Merriam-Webster Dictionary defines "mechanical" as relating to machinery or tools.

Power flow through a machine provides a way to understand the performance of devices ranging from levers and gear trains to automobiles and robotic systems. The German mechanician Franz Reuleaux wrote "a machine is a combination of resistant bodies so arranged that by their means the mechanical forces of nature can be compelled to do work accompanied by certain determinate motion." Notice that forces and motion combine to define power.

More recently, Uicker et al. stated that a machine is "a device for applying power or changing its direction." And McCarthy and Soh describe a machine as a system that "generally consists of a power source and a mechanism for the controlled use of this power."

Simple Machines

The idea that a machine can be decomposed into simple movable elements led Archimedes to define the lever, pulley and screw as simple machines. By the time of the Renaissance this list increased to include the wheel and axle, wedge and inclined plane. The modern approach to characterizing machines focusses on the components that allow movement, known as joints.

Wedge (hand axe): Perhaps the first example of a device designed to manage power is the hand axe. A hand axe is made by chipping stone, generally flint, to form a bifacial edge, or wedge. A wedge is a simple machine that transforms lateral force and movement of the tool into a transverse splitting force and movement of the workpiece. The available power is limited by the effort of the person using the tool, but because power is the product of force and movement, the wedge amplifies the force by reducing the movement. This amplification, or mechanical advantage is the ratio of the input speed to output speed. For a wedge this is given by $1/\tan\alpha$, where α is the tip angle. The faces of a wedge are modeled as straight lines to form a sliding or prismatic joint.

Lever: The lever is another important and simple device for managing power. This is a body that pivots on a fulcrum. Because the velocity of a point farther from the pivot is greater than the velocity of a point near the pivot, forces applied far from the pivot are amplified near the pivot by the associated decrease in speed. If a is the distance from the pivot to the point where the input force is applied and b is the distance to the point where the output force is applied, then a/b is the mechanical advantage of the lever. The fulcrum of a lever is modeled as a hinged or revolute joint.

Wheel: The wheel is clearly an important early machine, such as the chariot. A wheel uses the law of the lever to reduce the force needed to overcome friction when pulling a load. To see this notice that the friction associated with pulling a load on the ground is approximately the same as the friction in a simple bearing that supports the load on the axle of a wheel. However, the wheel forms a lever that magnifies the pulling force so that it overcomes the frictional resistance in the bearing.

Power Sources

Natural forces such as wind and water powered larger mechanical systems. Waterwheels appeared around the world around 300 BC to use flowing water to generate rotary motion, which was applied to milling grain, and powering lumber, machining and textile operations. Modern water turbines use water flowing through a dam to drive an electric generator. Early windmills captured wind power to generate rotary motion for milling operations. Modern wind turbines also drives a generator. This electricity in turn is used to drive motors forming the actuators of mechanical systems.

The word engine derives from "ingenuity" and originally referred to contrivances that may or may not be physical devices. A steam engine uses heat to boil water contained in a pressure vessel; the expanding steam drives a piston or a turbine. This principle can be seen in the aeolipile of Hero of Alexandria. This is called an external combustion engine.

Early Ganz Electric Generator in Zwevegem, West Flanders, Belgium

An automobile engine is called an internal combustion engine because it burns fuel (an exothermic chemical reaction) inside a cylinder and uses the expanding gases to drive a piston. A jet engine uses a turbine to compress air which is burned with fuel so that it expands through a nozzle to provide thrust to an aircraft, and so is also an "internal combustion engine."

The heat from coal and natural gas combustion in a boiler generates steam that drives a steam turbine to rotate an electric generator. A nuclear power plant uses heat from a nuclear reactor to generate steam and electric power. This power is distributed through a network of transmission lines for industrial and individual use. Electric motors use either AC or DC electric current to generate rotational movement. Electric servomotors are the actuators for mechanical systems ranging from robotic systems to modern aircraft. Hydraulic and pneumatic systems use electrically driven pumps to drive water or air respectively into cylinders to power linear movement.

Mechanisms

A machine consists of an actuator input, a system of mechanisms that generate the output forces and movement, and an interface to the user. Electric motors, hydraulic and pneumatic actuators provide the input forces and movement. This input is shaped by mechanisms consisting of gears and gear trains, belt and chain drives, cam and follower mechanisms, and linkages as well as friction devices such as brakes and clutches. Structural components consist of the frame, fasteners, bearings, springs, lubricants and seals, as well as a variety of specialized machine elements such as splines, pins and keys. The user interface ranges from switches and buttons to programmable logic controllers and includes the covers that provide texture, color and styling.

Gears and Gear Trains

The transmission of rotation between contacting toothed wheels can be traced back to the Antikythera mechanism of Greece and the south-pointing chariot of China. Illustrations by the renaissance scientist Georgius Agricola show gear trains with cylindrical teeth. The implementation of the involute tooth yielded a standard gear design that provides a constant speed ratio. Some important features of gears and gear trains are:

- The ratio of the pitch circles of mating gears defines the speed ratio and the mechanical advantage of the gear set.

- A planetary gear train provides high gear reduction in a compact package.

- It is possible to design gear teeth for gears that are non-circular, yet still transmit torque smoothly.

- The speed ratios of chain and belt drives are computed in the same way as gear ratios.

The Antikythera mechanism (main fragment)

Cam and Follower Mechanisms

A cam and follower is formed by the direct contact of two specially shaped links. The driving link is called the cam and the link that is driven through the direct contact of their surfaces is called the follower. The shape of the contacting surfaces of the cam and follower determines the movement of the mechanism.

Linkages

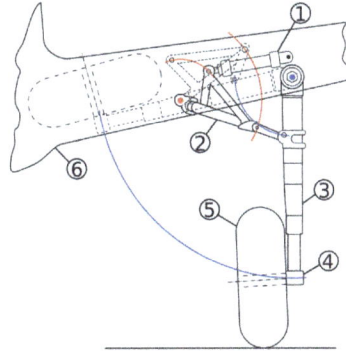

Schematic of the actuator and four-bar linkage that position an aircraft landing gear.

A linkage is a collection of links connected by joints. Generally, the links are the structural elements and the joints allow movement. Perhaps the single most useful example is the planar four-bar linkage. However, there are many more special linkages:

- Watt's linkage is a four-bar linkage that generates an approximate straight line. It was critical to the operation of his design for the steam engine. This linkage also appears in vehicle suspensions to prevent side-to-side movement of the body relative to the wheels.

- The success of Watt's linkage lead to the design of similar approximate straight-line linkages, such as Hoeken's linkage and Chebyshev's linkage.

- The Peaucellier linkage generates a true straight-line output from a rotary input.

- The Sarrus linkage is a spatial linkage that generates straight-line movement from a rotary input. Select this link for an animation of the Sarrus linkage

- The Klann linkage and the Jansen linkage are recent inventions that provide interesting walking movements. They are respectively a six-bar and an eight-bar linkage.

Flexure mechanisms

A flexure mechanism consists of a series of rigid bodies connected by compliant elements (also known as flexure joints) that is designed to produce a geometrically well-defined motion upon application of a force.

Structural Components

A number of machine elements provide important structural functions such as the frame, bearings, splines, spring and seals.

- The recognition that the frame of a mechanism is an important machine element changed the name three-bar linkage into four-bar linkage. Frames are generally assembled from truss or beam elements.

- Bearings are components designed to manage the interface between moving elements and are the source of friction in machines. In general, bearings are designed for pure rotation or straight line movement.

- Splines and keys are two ways to reliably mount an axle to a wheel, pulley or gear so that torque can be transferred through the connection.

- Springs provides forces that can either hold components of a machine in place or acts as a suspension to support part of a machine.

- Seals are used between mating parts of a machine to ensure fluids, such as water, hot gases, or lubricant do not leak between the mating surfaces.

- Fasteners such as screws, bolts, spring clips, and rivets are critical to the assembly of components of a machine. Fasteners are generally considered to be removable. In contrast, joining methods, such as welding, soldering, crimping and the application of adhesives, usually require cutting the parts to disassemble the components

Mechanics

Usher reports that Hero of Alexandria's treatise on *Mechanics* focussed on the study of lifting heavy weights. Today mechanics refers to the mathematical analysis of the forces and movement of a mechanical system, and consists of the study of the kinematics and dynamics of these systems.

Dynamics of Machines

The dynamic analysis of machines begins with a rigid-body model to determine reactions at the bearings, at which point the elasticity effects are included. The rigid-body dynamics studies the movement of systems of interconnected bodies under the action of external forces. The assumption that the bodies are rigid, which means that they do not deform under the action of applied forces, simplifies the analysis by reducing the parameters that describe the configuration of the system to the translation and rotation of reference frames attached to each body.

The dynamics of a rigid body system is defined by its equations of motion, which are derived using either Newtons laws of motion or Lagrangian mechanics. The solution of these equations of motion defines how the configuration of the system of rigid bodies changes as a function of time. The formulation and solution of rigid body dynamics is an important tool in the computer simulation of mechanical systems.

Kinematics of Machines

The dynamic analysis of a machine requires the determination of the movement, or kinematics, of its component parts, known as kinematic analysis. The assumption that the system is an assembly of rigid components allows rotational and translational movement to be modeled mathematically as Euclidean, or rigid, transformations. This allows the position, velocity and acceleration of all points in a component to be determined from these properties for a reference point, and the angular position, angular velocity and angular acceleration of the component.

Kinematic Chains

The classification of simple machines to provide a strategy for the design of new machines was developed by Franz Reuleaux, who collected and studied over 800 elementary machines. He recognized that the classical simple machines can be separated into the lever, pulley and wheel and axle that are formed by a body rotating about a hinge, and the inclined plane, wedge and screw that are similarly a block sliding on a flat surface.

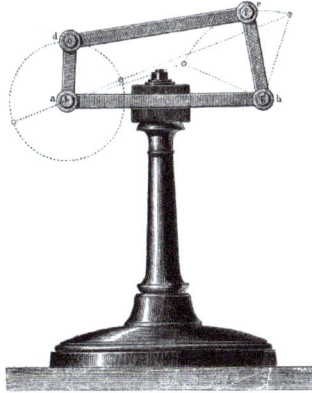

Illustration of a four-bar linkage from Kinematics of Machinery, 1876

Simple machines are elementary examples of kinematic chains or linkages that are used to model mechanical systems ranging from the steam engine to robot manipulators. The bearings that form the fulcrum of a lever and that allow the wheel and axle and pulleys to rotate are examples of a kinematic pair called a hinged joint. Similarly, the flat surface of an inclined plane and wedge are examples of the kinematic pair called a sliding joint. The screw is usually identified as its own kinematic pair called a helical joint.

This realization shows that it is the joints, or the connections that provide movement, that are the primary elements of a machine. Starting with four types of joints, the rotary joint, sliding joint, cam joint and gear joint, and related connections such as cables and belts, it is possible to understand a machine as an assembly of solid parts that connect these joints called a mechanism .

Two levers, or cranks, are combined into a planar four-bar linkage by attaching a link that connects the output of one crank to the input of another. Additional links can be attached to form a six-bar linkage or in series to form a robot.

Planar Mechanisms

While all mechanisms in a mechanical system are three-dimensional, they can be analyzed using plane geometry, if the movement of the individual components are constrained so all point trajectories are parallel or in a series connection to a plane. In this case the system is called a *planar mechanism*. The kinematic analysis of planar mechanisms uses the subset of SE(3) consisting of planar rotations and translations, denote SE(2).

The group SE(2) is three-dimensional, which means that every position of a body in the plane is defined by three parameters. The parameters are often the x and y coordinates of the origin of a coordinate frame in M measured from the origin of a coordinate frame in F, and the angle measured from the x-axis in F to the x-axis in M. This is often described saying a body in the plane has three degrees-of-freedom.

The pure rotation of a hinge and the linear translation of a slider can be identified with subgroups of SE(2), and define the two joints one degree-of-freedom joints of planar mechanisms. The cam joint formed by two surfaces in sliding and rotating contact is a two degree-of-freedom joint.

Spherical Mechanisms

It is possible to construct a mechanism such that the point trajectories in all components lie in concentric spherical shells around a fixed point. An example is the gimbaled gyroscope. These devices are called *spherical mechanisms*. Spherical mechanisms are constructed by connecting links with hinged joints such that the axes of each hinge passes through the same point. This point becomes center of the concentric spherical shells. The movement of these mechanisms is characterized by the group SO(3) of rotations in three-dimensional space. Other examples of spherical mechanisms are the automotive differential and the robotic wrist.

Select this link for an animation of a Spherical deployable mechanism.

The rotation group SO(3) is three-dimensional. An example of the three parameters that specify a spatial rotation are the roll, pitch and yaw angles used to define the orientation of an aircraft.

Spatial Mechanisms

A mechanism in which a body moves through a general spatial movement is called a *spatial mechanism*. An example is the RSSR linkage, which can be viewed as a four-bar linkage in which the hinged joints of the coupler link are replaced by rod ends, also called spherical joints or ball joints. The rod ends allow the input and output cranks of the RSSR linkage to be misaligned to the point that they lie in different planes, which causes the coupler link to move in a general spatial movement. Robot arms, Stewart platforms, and humanoid robotic systems are also examples of spatial mechanisms.

Bennett's linkage is an example of a spatial overconstrained mechanism, which is constructed from four hinged joints.

The group SE(3) is six-dimensional, which means the position of a body in space is defined by six parameters. Three of the parameters define the origin of the moving reference frame relative to the fixed frame. Three other parameters define the orientation of the moving frame relative to the fixed frame.

Kinematic Diagram

A kinematic diagram reduces the machine components to a skeleton diagram that emphasizes the joints and reduces the links to simple geometric elements. This diagram can also be formulated as a graph by representing the links of the mechanism as vertices and the joints as edges of the graph. This version of the kinematic diagram has proven effective in enumerating kinematic structures in the process of machine design.

An important consideration in this design process is the degree of freedom of the system of links and joints, which is be determined using the Chebychev–Grübler–Kutzbach criterion.

Machine Design

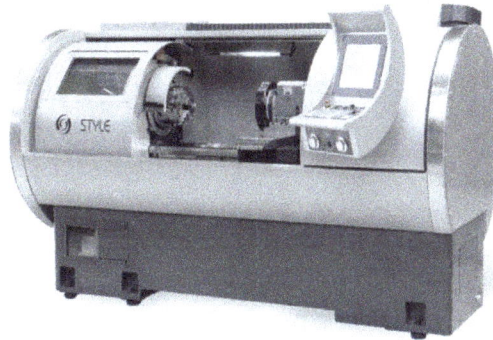

A CNC metalworking lathe.

Machine design refers to the procedures and techniques used to address the three phases of a machine's lifecycle:

1. invention, which involves the identification of a need, development of requirements, concept generation, prototype development, manufacturing, and verification testing;

2. performance engineering involves enhancing manufacturing efficiency, reducing service and maintenance demands, adding features and improving effectiveness, and validation testing;

3. recycle is the decommissioning and disposal phase and includes recovery and reuse of materials and components.

Machine Elements

The elementary mechanical components of a machine are termed machine elements. These elements consist of three basic types (i) *structural components* such as frame members, bearings, axles, splines, fasteners, seals, and lubricants, (ii) *mechanisms* that control movement in various ways such as gear trains, belt or chain drives, linkages, cam and follower systems, including brakes and clutches, and (iii) *control components* such as buttons, switches, indicators, sensors, actuators and computer controllers. While generally not considered to be a machine element, the shape, texture and color of covers are an important part of a machine that provide a styling and operational interface between the mechanical components of a machine and its users.

Mechanical System

A mechanical system manages power to accomplish a task that involves forces and movement.

The Oxford English Dictionary defines the adjective *mechanical* as skilled in the practical application of an art or science, of the nature of a machine or machines, and relating to or caused by movement, physical forces, properties or agents such as is dealt with by Mechanics. Similarly Merriam-Webster Dictionary defines "mechanical" as relating to machinery or tools.

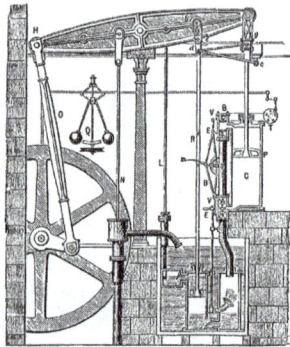

The Boulton & Watt Steam Engine, 1784.

A mechanical system consists of (i) a power source and actuators that generate forces and movement, (ii) a system of mechanisms that shape the actuator input to achieve a specific application of output forces and movement, and (iii) a controller with sensors that compares the output to a performance goal and then directs the actuator input. This can be seen in Watt's steam engine in which the power is provided by steam expanding to drive the piston. The walking beam, coupler and crank transform the linear movement of the piston into rotation of the output pulley. Finally, the pulley rotation drives the flyball governor which controls the valve for the steam input to the piston cylinder.

Power flow through a mechanical system provides a way to understand the performance of devices ranging from levers and gear trains to automobiles and robotic systems.

Power Sources

Human and animal effort were the original power sources for early machines. Natural forces such as wind and water powered larger mechanical systems.

Waterwheel: Waterwheels appeared around the world around 300 BC to use flowing water to generate rotary motion, which was applied to milling grain, and powering lumber, machining and textile operations. Modern water turbines use water flowing through a dam to drive an electric generator.

Windmill: Early windmills captured wind power to generate rotary motion for milling operations. Modern wind turbines also drives a generator. This electricity in turn is used to drive motors forming the actuators of mechanical systems.

Engine: The word engine derives from "ingenuity" and originally referred to contrivances that may or may not be physical devices. See Merriam-Webster's definition of engine. A steam engine uses heat to boil water contained in a pressure vessel; the expanding steam drives a piston or a turbine. This principle can be seen in the aeolipile of Hero of Alexandria. This is called an external combustion engine.

An automobile engine is called an internal combustion engine because it burns fuel (an exothermic chemical reaction) inside a cylinder and uses the expanding gases to drive a piston. A jet engine uses a turbine to compress air which is burned with fuel so that it expands through a nozzle to provide thrust to an aircraft, and so is also an "internal combustion engine."

Power plant: The heat from coal and natural gas combustion in a boiler generates steam that drives a steam turbine to rotate an electric generator. A nuclear power plant uses heat from a nuclear reactor to generate steam and electric power. This power is distributed through a network of transmission lines for industrial and individual use.

Motors: Electric motors use either AC or DC electric current to generate rotational movement. Electric servomotors are the actuators for mechanical systems ranging from robotic systems to modern aircraft.

Fluid Power: Hydraulic and pneumatic systems use electrically driven pumps to drive water or air respectively into cylinders to power linear movement.

Mechanisms

The *mechanism* of a mechanical system is assembled from components called *machine elements*. These elements provide structure for the system and control its movement.

The structural components are, generally, the frame members, bearings, splines, springs, seals, fasteners and covers. The shape, texture and color of covers provide a styling and operational interface between the mechanical system and its users.

The assemblies that control movement are also called "mechanisms." Mechanisms are generally classified as gears and gear trains, which includes belt drives and chain drives, cam and follower mechanisms, and linkages, though there are other special mechanisms such as clamping linkages, indexing mechanisms, escapements and friction devices such as brakes and clutches.

The number of degrees of freedom (dof) or mobility of a mechanism depends on the number of links and joints and the types of joints used to construct the mechanism. The general mobility of a mechanism is the difference between the unconstrained freedom of the links and the number of constraints imposed by the joints. It is described by the Chebychev-Grübler-Kutzbach criterion.

Controllers

Controllers combine sensors, logic, and actuators to maintain the performance of components of a machine. Perhaps the best known is the flyball governor for a steam engine. Examples of these devices range from a thermostat that as temperature rises opens a valve to cooling water to speed controllers such the cruise control system in an automobile. The programmable logic controller replaced relays and specialized control mechanisms with a programmable computer. Servomotors that accurately position a shaft in response to an electrical command are the actuators that make robotic systems possible.

References

- Al-Jazarí. The Book of Knowledge of Ingenious Mechanical Devices: Kitáb fí ma'rifat al-hiyal al-handasiyya. Springer, 1973. ISBN 90-277-0329-9.

- Asimov, Isaac (1988), Understanding Physics, New York, New York, USA: Barnes & Noble, p. 88, ISBN 0-88029-251-2.

- Chiu, Y. C. (2010), An introduction to the History of Project Management, Delft: Eburon Academic Publishers, p. 42, ISBN 90-5972-437-2

- Usher, Abbott Payson (1988). A History of Mechanical Inventions. USA: Courier Dover Publications. p. 98. ISBN 0-486-25593-X.

- Krebs, Robert E. (2004). Groundbreaking Experiments, Inventions, and Discoveries of the Middle Ages. Greenwood Publishing Group. p. 163. ISBN 0-313-32433-6. Retrieved 2008-05-21.

- Stephen, Donald; Lowell Cardwell (2001). Wheels, clocks, and rockets: a history of technology. USA: W. W. Norton & Company. pp. 85–87. ISBN 0-393-32175-4.

- Occupational Employment and Wages, 17-2141 Mechanical Engineers. U.S. Bureau of Labor, May 2012. Accessed: 15 February 2014.

- engineering "mechanical engineering". The American Heritage Dictionary of the English Language, Fourth Edition. Retrieved: 19 September 2014.

Concepts of Mechanical Engineering

Mechanics, kinematics and structural analysis are some significant topics related to mechanical engineering. Mechanics is an area of science that deals with the behaviour of physical bodies. The following chapter unfolds the main concepts and aspects of mechanical engineering in a critical manner.

Mechanics

Mechanics is an area of science concerned with the behaviour of physical bodies when subjected to forces or displacements, and the subsequent effects of the bodies on their environment. The scientific discipline has its origins in Ancient Greece with the writings of Aristotle and Archimedes. During the early modern period, scientists such as Khayaam, Galileo, Kepler, and Newton, laid the foundation for what is now known as classical mechanics. It is a branch of classical physics that deals with particles that are either at rest or are moving with velocities significantly less than the speed of light. It can also be defined as a branch of science which deals with the motion of and forces on objects.

Classical Versus Quantum Mechanics

Historically, classical mechanics came first, while quantum mechanics is a comparatively recent invention. Classical mechanics originated with Isaac Newton's laws of motion in *Principia Mathematica*; Quantum Mechanics was discovered in the early 20th century. Both are commonly held to constitute the most certain knowledge that exists about physical nature. Classical mechanics has especially often been viewed as a model for other so-called exact sciences. Essential in this respect is the relentless use of mathematics in theories, as well as the decisive role played by experiment in generating and testing them.

Quantum mechanics is of a wider scope, as it encompasses classical mechanics as a sub-discipline which applies under certain restricted circumstances. According to the correspondence principle, there is no contradiction or conflict between the two subjects, each simply pertains to specific situations. The correspondence principle states that the behavior of systems described by quantum theories reproduces classical physics in the limit of large quantum numbers. Quantum mechanics has superseded classical mechanics at the foundational level and is indispensable for the explanation and prediction of processes at the molecular, atomic, and sub-atomic level. However, for macroscopic processes classical mechanics is able to solve problems which are unmanageably difficult in quantum mechanics and hence remains useful and well used. Modern descriptions of such behavior begin with a careful definition of such quantities as displacement (distance moved), time, velocity, acceleration, mass, and force. Until about 400 years ago, however, motion was explained from a very different point of view. For example, following the ideas of Greek philosopher and scientist Aristotle, scientists reasoned that a cannonball falls down because its natural position is in the Earth; the sun, the moon, and the stars travel in circles around the earth because it is the nature of heavenly objects to travel in perfect circles.

Often cited as the father of modern science, Galileo brought together the ideas of other great thinkers of his time and began to analyze motion in terms of distance traveled from some starting position and the time that it took. He showed that the speed of falling objects increases steadily during the time of their fall. This acceleration is the same for heavy objects as for light ones, provided air friction (air resistance) is discounted. The English mathematician and physicist Isaac Newton improved this analysis by defining force and mass and relating these to acceleration. For objects traveling at speeds close to the speed of light, Newton's laws were superseded by Albert Einstein's theory of relativity. For atomic and subatomic particles, Newton's laws were superseded by quantum theory. For everyday phenomena, however, Newton's three laws of motion remain the cornerstone of dynamics, which is the study of what causes motion.

Relativistic Versus Newtonian Mechanics

In analogy to the distinction between quantum and classical mechanics, Einstein's general and special theories of relativity have expanded the scope of Newton and Galileo's formulation of mechanics. The differences between relativistic and Newtonian mechanics become significant and even dominant as the velocity of a massive body approaches the speed of light. For instance, in Newtonian mechanics, Newton's laws of motion specify that $F = ma$, whereas in Relativistic mechanics and Lorentz transformations, which were first discovered by Hendrik Lorentz, $F = \gamma ma$ (where γ is the Lorentz factor, which is almost equal to 1 for low speeds).

General Relativistic Versus Quantum

Relativistic corrections are also needed for quantum mechanics, although general relativity has not been integrated. The two theories remain incompatible, a hurdle which must be overcome in developing a theory of everything.

Antiquity

The main theory of mechanics in antiquity was Aristotelian mechanics. A later developer in this tradition is Hipparchus.

Medieval Age

In the Middle Ages, Aristotle's theories were criticized and modified by a number of figures, beginning with John Philoponus in the 6th century. A central problem was that of projectile motion, which was discussed by Hipparchus and Philoponus. This led to the development of the theory of impetus by 14th-century French priest Jean Buridan, which developed into the modern theories of inertia, velocity, acceleration and momentum. This work and others was developed in 14th-century England by the Oxford Calculators such as Thomas Bradwardine, who studied and formulated various laws regarding falling bodies.

Arabic Machine Manuscript. Unknown date (at a guess: 16th to 19th centuries).

On the question of a body subject to a constant (uniform) force, the 12th-century Jewish-Arab Nathanel (Iraqi, of Baghdad) stated that constant force imparts constant acceleration, while the main properties are uniformly accelerated motion (as of falling bodies) was worked out by the 14th-century Oxford Calculators.

Early Modern Age

Two central figures in the early modern age are Galileo Galilei and Isaac Newton. Galileo's final statement of his mechanics, particularly of falling bodies, is his *Two New Sciences* (1638). Newton's 1687 *Philosophiæ Naturalis Principia Mathematica* provided a detailed mathematical account of mechanics, using the newly developed mathematics of calculus and providing the basis of Newtonian mechanics.

There is some dispute over priority of various ideas: Newton's *Principia* is certainly the seminal work and has been tremendously influential, and the systematic mathematics therein did not and could not have been stated earlier because calculus had not been developed. However, many of the ideas, particularly as pertain to inertia (impetus) and falling bodies had been developed and stated by earlier researchers, both the then-recent Galileo and the less-known medieval predecessors. Precise credit is at times difficult or contentious because scientific language and standards of proof changed, so whether medieval statements are *equivalent* to modern statements or *sufficient* proof, or instead *similar* to modern statements and *hypotheses* is often debatable.

Modern Age

Two main modern developments in mechanics are general relativity of Einstein, and quantum mechanics, both developed in the 20th century based in part on earlier 19th-century ideas. The development in the modern continuum mechanics, particularly in the areas of elasticity, plasticity, fluid dynamics, electrodynamics and thermodynamics of deformable media, started in the second half of the 20th century.

Types of Mechanical Bodies

The often-used term body needs to stand for a wide assortment of objects, including particles, projectiles, spacecraft, stars, parts of machinery, parts of solids, parts of fluids (gases and liquids), etc.

Other distinctions between the various sub-disciplines of mechanics, concern the nature of the bodies being described. Particles are bodies with little (known) internal structure, treated as mathematical points in classical mechanics. Rigid bodies have size and shape, but retain a simplicity close to that of the particle, adding just a few so-called degrees of freedom, such as orientation in space.

Otherwise, bodies may be semi-rigid, i.e. elastic, or non-rigid, i.e. fluid. These subjects have both classical and quantum divisions of study.

For instance, the motion of a spacecraft, regarding its orbit and attitude (rotation), is described by the relativistic theory of classical mechanics, while the analogous movements of an atomic nucleus are described by quantum mechanics.

Sub-Disciplines in Mechanics

The following are two lists of various subjects that are studied in mechanics.

Note that there is also the "theory of fields" which constitutes a separate discipline in physics, formally treated as distinct from mechanics, whether classical fields or quantum fields. But in actual practice, subjects belonging to mechanics and fields are closely interwoven. Thus, for instance, forces that act on particles are frequently derived from fields (electromagnetic or gravitational), and particles generate fields by acting as sources. In fact, in quantum mechanics, particles themselves are fields, as described theoretically by the wave function.

Classical Mechanics

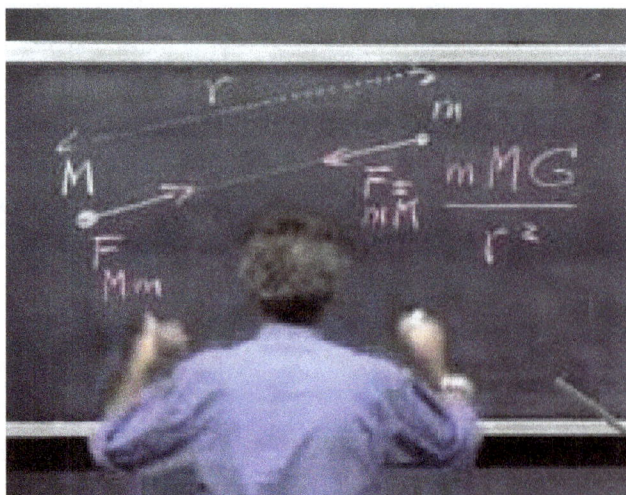

Prof. Walter Lewin explains Newton's law of gravitation in MIT course 8.01

The following are described as forming classical mechanics:

- Newtonian mechanics, the original theory of motion (kinematics) and forces (dynamics).
- Analytical mechanics is a reformulation of Newtonian mechanics with an emphasis on system energy, rather than on forces. There are two main branches of analytical mechanics:
    - Hamiltonian mechanics, a theoretical formalism, based on the principle of conservation of energy.
    - Lagrangian mechanics, another theoretical formalism, based on the principle of the least action.
- Classical statistical mechanics generalizes ordinary classical mechanics to consider systems in an unknown state; often used to derive thermodynamic properties.
- Celestial mechanics, the motion of bodies in space: planets, comets, stars, galaxies, etc.
- Astrodynamics, spacecraft navigation, etc.
- Solid mechanics, elasticity, the properties of deformable bodies.
- Fracture mechanics
- Acoustics, sound (= density variation propagation) in solids, fluids and gases.
- Statics, semi-rigid bodies in mechanical equilibrium
- Fluid mechanics, the motion of fluids
- Soil mechanics, mechanical behavior of soils
- Continuum mechanics, mechanics of continua (both solid and fluid)
- Hydraulics, mechanical properties of liquids
- Fluid statics, liquids in equilibrium
- Applied mechanics, or Engineering mechanics
- Biomechanics, solids, fluids, etc. in biology
- Biophysics, physical processes in living organisms
- Relativistic or Einsteinian mechanics, universal gravitation.

Quantum Mechanics

The following are categorized as being part of quantum mechanics:

- Schrödinger wave mechanics, used to describe the motion of the wavefunction of a single particle.
- Matrix mechanics is an alternative formulation that allows considering systems with a finite-dimensional state space.
- Quantum statistical mechanics generalizes ordinary quantum mechanics to consider sys-

tems in an unknown state; often used to derive thermodynamic properties.

- Particle physics, the motion, structure, and reactions of particles
- Nuclear physics, the motion, structure, and reactions of nuclei
- Condensed matter physics, quantum gases, solids, liquids, etc.

Kinematics

Kinematics is the branch of classical mechanics which describes the motion of points (alternatively "particles"), bodies (objects), and systems of bodies without consideration of the masses of those objects nor the forces that may have caused the motion. Kinematics as a field of study is often referred to as the "geometry of motion" and as such may be seen as a branch of mathematics. Kinematics begins with a description of the geometry of the system and the initial conditions of known values of the position, velocity and or acceleration of various points that are a part of the system, then from geometrical arguments it can determine the position, the velocity and the acceleration of any part of the system. The study of the influence of forces acting on masses falls within the purview of kinetics.

Kinematics is used in astrophysics to describe the motion of celestial bodies and collections of such bodies. In mechanical engineering, robotics, and biomechanics kinematics is used to describe the motion of systems composed of joined parts (multi-link systems) such as an engine, a robotic arm or the skeleton of the human body.

The use of geometric transformations, also called rigid transformations, to describe the movement of components of a mechanical system simplifies the derivation of its equations of motion, and is central to dynamic analysis.

Kinematic analysis is the process of measuring the kinematic quantities used to describe motion. In engineering, for instance, kinematic analysis may be used to find the range of movement for a given mechanism, and working in reverse, using kinematic synthesis used to design a mechanism for a desired range of motion. In addition, kinematics applies algebraic geometry to the study of the mechanical advantage of a mechanical system or mechanism.

Etymology of The Term

Kinematic and cinématique are related to the French word cinéma, but neither are directly derived from it. However, they do share a root word in common, as cinéma came from the shortened form of cinématographe, "motion picture projector and camera," once again from the Greek word for movement but also the Greek word for writing.

Kinematics of a Particle Trajectory in a Non-Rotating Frame of Reference

Kinematic vectors in plane polar coordinates. Notice the setup is not restricted to 2d space, but a plane in any higher dimension.

$$\mathbf{a} = \frac{d\mathbf{v}}{dt}$$

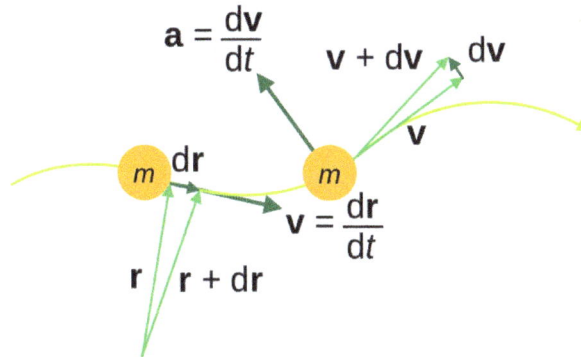

Kinematic quantities of a classical particle: mass m, position $\mathbf{r}$, velocity $\mathbf{v}$, acceleration $\mathbf{a}$.

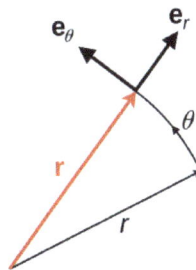

Position vector $\mathbf{r}$, always points radially from the origin.

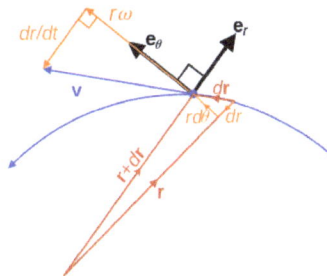

Velocity vector v, always tangent to the path of motion.

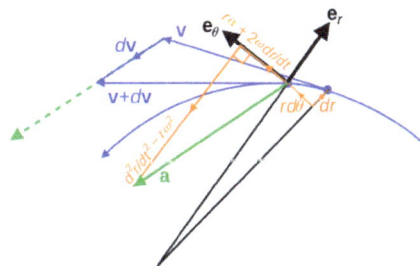

Acceleration vector **a**, not parallel to the radial motion but offset by the angular and Coriolis accelerations, nor tangent to the path but offset by the centripetal and radial accelerations.

Particle kinematics is the study of the trajectory of a particle. The position of a particle is defined to be the coordinate vector from the origin of a coordinate frame to the particle. For example, consider a tower 50 m south from your home, where the coordinate frame is located at your home, such that East is the x-direction and North is the y-direction, then the coordinate vector to the base of the tower is r=(0, -50, 0). If the tower is 50 m high, then the coordinate vector to the top of the tower is r=(0, -50, 50).

In the most general case, a three-dimensional coordinate system is used to define the position of a particle. However, if the particle is constrained to move in a surface, a two-dimensional coordinate system is sufficient. All observations in physics are incomplete without those observations being described with respect to a reference frame.

The position vector of a particle is a vector drawn from the origin of the reference frame to the particle. It expresses both the distance of the point from the origin and its direction from the origin. In three dimensions, the position of point P can be expressed as

$$\mathbf{P} = (x_P, y_P, z_P) = x_P\hat{i} + y_P\hat{j} + z_P\hat{k},$$

where x_p, y_p, and z_p are the Cartesian coordinates and i, j and k are the unit vectors along the x, y, and z coordinate axes, respectively. The magnitude of the position vector $|P|$ gives the distance between the point P and the origin.

$$|\mathbf{P}| = \sqrt{x_P^2 + y_P^2 + z_P^2}.$$

The direction cosines of the position vector provide a quantitative measure of direction. It is important to note that the position vector of a particle isn't unique. The position vector of a given particle is different relative to different frames of reference.

The *trajectory* of a particle is a vector function of time, P(t), which defines the curve traced by the moving particle, given by

$$\mathbf{P}(t) = x_P(t)\hat{i} + y_P(t)\hat{j} + z_P(t)\hat{k},$$

where the coordinates x_p, y_p, and z_p are each functions of time.

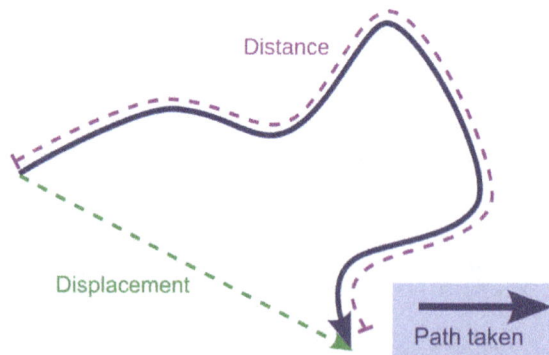

The distance travelled is always greater than or equal to the displacement.

Velocity and Speed

The velocity of a particle is a vector quantity that describes the direction of motion and the magnitude of the motion of particle. More mathematically, the rate of change of the position vector of a point, with respect to time is the velocity of the point. Consider the ratio of the difference of two positions of a particle divided by the time interval, which is called the average velocity over that time interval. This average velocity is defined as

$$\overline{\mathbf{V}} = \frac{\Delta \mathbf{P}}{\Delta t},$$

where ΔP is the change in the position vector over the time interval Δt.

In the limit as the time interval Δt becomes smaller and smaller, the average velocity becomes the time derivative of the position vector,

$$\mathbf{V} = \lim_{\Delta t \to 0} \frac{\Delta \mathbf{P}}{\Delta t} = \frac{d\mathbf{P}}{dt} = \dot{\mathbf{P}} = \dot{x}_p\hat{i} + \dot{y}_P\hat{j} + \dot{z}_p\hat{k}.$$

Thus, velocity is the time rate of change of position of a point, and the dot denotes the derivative of those functions x, y, and z with respect to time. Furthermore, the velocity is tangent to the trajectory of the particle at every position the particle occupies along its path. Note that in a non-rotating frame of reference, the derivatives of the coordinate directions are not considered as their directions and magnitudes are constants.

The speed of an object is the magnitude |V| of its velocity. It is a scalar quantity:

$$|\mathbf{V}| = |\dot{\mathbf{P}}| = \frac{ds}{dt},$$

where s is the arc-length measured along the trajectory of the particle. This arc-length traveled by a particle over time is a non-decreasing quantity. Hence, ds/dt is non-negative, which implies that speed is also non-negative.

Acceleration

The velocity vector can change in magnitude and in direction or both at once. Hence, the acceleration is the rate of change of the magnitude of the velocity vector plus the rate of change of direction of that vector. The same reasoning used with respect to the position of a particle to define velocity, can be applied to the velocity to define acceleration. The acceleration of a particle is the vector defined by the rate of change of the velocity vector. The average acceleration of a particle over a time interval is defined as the ratio.

$$\overline{\mathbf{A}} = \frac{\Delta \mathbf{V}}{\Delta t},$$

where ΔV is the difference in the velocity vector and Δt is the time interval.

The acceleration of the particle is the limit of the average acceleration as the time interval approaches zero, which is the time derivative,

$$\text{Eqn 1) } \mathbf{A} = \lim_{\Delta t \to 0} \frac{\Delta \mathbf{V}}{\Delta t} = \frac{d\mathbf{V}}{dt} = \dot{\mathbf{V}} = \dot{v}_x\hat{i} + \dot{v}_y\hat{j} + \dot{v}_z\hat{k}$$

or

$$\mathbf{A} = \ddot{\mathbf{P}} = \ddot{x}_p\hat{i} + \ddot{y}_P\hat{j} + \ddot{z}_P\hat{k}$$

Thus, acceleration is the first derivative of the velocity vector and the second derivative of the position vector of that particle. Note that in a non-rotating frame of reference, the derivatives of the coordinate directions are not considered as their directions and magnitudes are constants.

The magnitude of the acceleration of an object is the magnitude |A| of its acceleration vector. It is a scalar quantity:

$$|\mathbf{A}| = |\dot{\mathbf{V}}| = \frac{dv}{dt},$$

Relative Position Vector

A relative position vector is a vector that defines the position of one point relative to another. It is the difference in position of the two points. The position of one point A relative to another point B is simply the difference between their positions

$$\mathbf{P}_{A/B} = \mathbf{P}_A - \mathbf{P}_B$$

which is the difference between the components of their position vectors.

If point A has position components

$$\mathbf{P}_A = \left(X_A, Y_A, Z_A \right)$$

If point B has position components

$$\mathbf{P}_B = \left(X_B, Y_B, Z_B \right)$$

then the position of point A relative to point B is the difference between their components:

$$\mathbf{P}_{A/B} = \mathbf{P}_A - \mathbf{P}_B = \left(X_A - X_B, Y_A - Y_B, Z_A - Z_B \right)$$

Relative Velocity

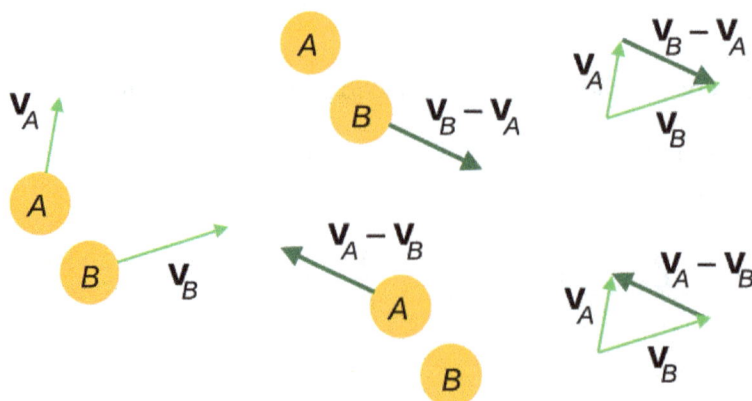

Relative velocities between two particles in classical mechanics.

The velocity of one point relative to another is simply the difference between their velocities

$$\mathbf{V}_{A/B} = \mathbf{V}_A - \mathbf{V}_B$$

which is the difference between the components of their velocities.

If point A has velocity components

$$\mathbf{V}_A = \left(V_{A_x}, V_{A_y}, V_{A_z} \right)$$

and point B has velocity components

$$\mathbf{V}_B = \left(V_{B_x}, V_{B_y}, V_{B_z} \right)$$

then the velocity of point A relative to point B is the difference between their components:

$$\mathbf{V}_{A/B} = \mathbf{V}_A - \mathbf{V}_B = \left(V_{A_x} - V_{B_x}, V_{A_y} - V_{B_y}, V_{A_z} - V_{B_z} \right)$$

Alternatively, this same result could be obtained by computing the time derivative of the relative position vector $\mathbf{R}_{B/A}$.

In the case where the velocity is close to the speed of light c (generally within 95%), another scheme of relative velocity called rapidity, that depends on the ratio of V to c, is used in special relativity.

Relative Acceleration

The acceleration of one point C relative to another point B is simply the difference between their accelerations.

$$\mathbf{A}_{C/B} = \mathbf{A}_C - \mathbf{A}_B$$

which is the difference between the components of their accelerations.

If point C has acceleration components

$$\mathbf{A}_C = \left(A_{C_x}, A_{C_y}, A_{C_z} \right)$$

and point B has acceleration components

$$\mathbf{A}_B = \left(A_{B_x}, A_{B_y}, A_{B_z} \right)$$

then the acceleration of point C relative to point B is the difference between their components:

$$\mathbf{A}_{C/B} = \mathbf{A}_C - \mathbf{A}_B = \left(A_{C_x} - A_{B_x}, A_{C_y} - A_{B_y}, A_{C_z} - A_{B_z} \right)$$

Alternatively, this same result could be obtained by computing the second time derivative of the relative position vector $\mathbf{P}_{B/A}$.

Particle Trajectories Under Constant Acceleration

For the case of constant acceleration, the differential equation Eq 1) can be integrated as the acceleration vector A of a point P is constant in magnitude and direction. Such a point is said to be undergoing *uniformly accelerated motion*. In this case, the velocity V(t) and then the trajectory P(t) of the particle can be obtained by integrating the acceleration equation A with respect to time.

Assuming that the initial conditions of the position, $\mathbf{P}_0$, and velocity $\mathbf{V}_0$ are known, the first integration yields the velocity of the particle as a function of time.

$$\mathbf{V}(t) = \int_0^t \mathbf{A}dt = \mathbf{A}t + \mathbf{V}_0.$$

A second integration yields its path (trajectory),

$$\mathbf{P}(t) = \int_0^t \mathbf{V}(t)dt = \int (\mathbf{A}t + \mathbf{V}_0)dt = \tfrac{1}{2}\mathbf{A}t^2 + \mathbf{V}_0 t + \mathbf{P}_0.$$

Additional relations between displacement, velocity, acceleration, and time can be derived. Since the acceleration is constant,

$\mathbf{A} = \dfrac{\Delta\mathbf{V}}{\Delta t} = \dfrac{\mathbf{V} - \mathbf{V}_0}{t}$ can be substituted into the above equation to give:

$$\mathbf{P}(t) = \mathbf{P}_0 + \left(\frac{\mathbf{V} + \mathbf{V}_0}{2}\right)t.$$

A relationship between velocity, position and acceleration without explicit time dependence can be had by solving the average acceleration for time and substituting and simplifying

$$t = \frac{\mathbf{V} - \mathbf{V}_0}{\mathbf{A}}$$

$$(\mathbf{P} - \mathbf{P}_0)\circ\mathbf{A} = (\mathbf{V} - \mathbf{V}_0)\circ\frac{\mathbf{V} + \mathbf{V}_0}{2},$$

where ∘ denotes the dot product, which is appropriate as the products are scalars rather than vectors.

$$2(\mathbf{P} - \mathbf{P}_0)\circ\mathbf{A} = |\mathbf{V}|^2 - |\mathbf{V}_0|^2.$$

The dot can be replaced by the cosine of the angle ∝ between the vectors and the vectors by their magnitudes, in which case:

$$2(|\mathbf{P}| - |\mathbf{P}_0|)|\mathbf{A}|cosine(\alpha) = |\mathbf{V}|^2 - |\mathbf{V}_0|^2.$$

In the case of acceleration always in the direction of the motion ∝ = 0, cosine(0) = 1 and,

$$|\mathbf{V}|^2 = |\mathbf{V}_0|^2 + 2|\mathbf{A}|(|\mathbf{P}| - |\mathbf{P}_0|).$$

This can be simplified using the notation for the magnitudes of the vectors $|\mathbf{A}| = a, |\mathbf{V}| = v, |\mathbf{P} - \mathbf{P}_0| = \delta s$ where δs can be any curvaceous path taken as the constant tangential acceleration is applied along that path, so

$$v^2 = v_0^2 + 2a(\delta s).$$

This reduces the parametric equations of motion of the particle to a cartesian relationship of speed versus position. This relation is useful when time is not known explicitly.

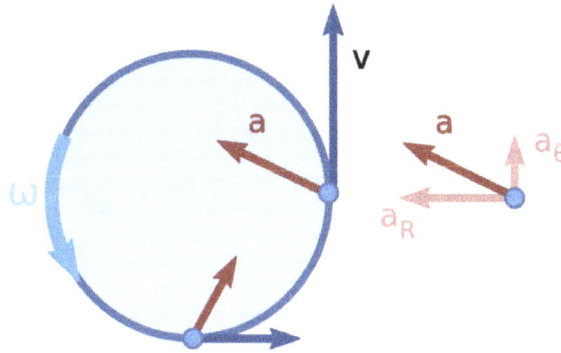

Figure: Velocity and acceleration for nonuniform circular motion: the velocity vector is tangential to the orbit, but the acceleration vector is not radially inward because of its tangential component a_θ that increases the rate of rotation: $d\omega/dt = |a_\theta|/R$.

Particle Trajectories in Cylindrical-Polar Coordinates

It is often convenient to formulate the trajectory of a particle P(t) = (X(t), Y(t) and Z(t)) using polar coordinates in the X-Y plane. In this case, its velocity and acceleration take a convenient form.

Recall that the trajectory of a particle P is defined by its coordinate vector P measured in a fixed reference frame F. As the particle moves, its coordinate vector P(t) traces its trajectory, which is a curve in space, given by:

$$\mathbf{P}(t) = X(t)\hat{i} + Y(t)\hat{j} + Z(t)\hat{k},$$

where i, j, and k are the unit vectors along the X, Y and Z axes of the reference frame F, respectively.

Consider a particle P that moves only on the surface of a circular cylinder R(t)=constant, it is possible to align the Z axis of the fixed frame F with the axis of the cylinder. Then, the angle θ around this axis in the X-Y plane can be used to define the trajectory as,

$$\mathbf{P}(t) = R\cos\theta(t)\hat{i} + R\sin\theta(t)\hat{j} + Z(t)\hat{k}.$$

The cylindrical coordinates for P(t) can be simplified by introducing the radial and tangential unit vectors,

$$\mathbf{e}_r = \cos\theta(t)\hat{i} + \sin\theta(t)\hat{j}, \quad \mathbf{e}_t = -\sin\theta(t)\hat{i} + \cos\theta(t)\hat{j}.$$

and their time derivatives from elementary calculus:

$$\frac{d}{dt}\mathbf{e}_r = \dot{\mathbf{e}}_r = \dot{\theta}\mathbf{e}_t$$

$$\frac{d}{dt}\dot{\mathbf{e}}_r = \ddot{\mathbf{e}}_r = \ddot{\theta}\mathbf{e}_t - \dot{\theta}\dot{\mathbf{e}}_r$$

$$\frac{d}{dt}\mathbf{e}_t = \dot{\mathbf{e}}_t = -\dot{\theta}\mathbf{e}_r$$

$$\frac{d}{dt}\dot{\mathbf{e}}_t = \ddot{\mathbf{e}}_t = -\ddot{\theta}\mathbf{e}_r - \dot{\theta}^2\mathbf{e}_t .$$

Using this notation, P(t) takes the form,

$$\mathbf{P}(t) = R\mathbf{e}_r + Z(t)\hat{k},$$

where R is constant in the case of the particle moving only on the surface of a cylinder of radius R.

In general, the trajectory P(t) is not constrained to lie on a circular cylinder, so the radius R varies with time and the trajectory of the particle in cylindrical-polar coordinates becomes:

$$\mathbf{P}(t) = R(t)\mathbf{e}_r + Z(t)\hat{k}.$$

Where R, theta, and Z might be continuously differentiable functions of time and the function notation is dropped for simplicity. The velocity vector V_p is the time derivative of the trajectory P(t), which yields:

$$\mathbf{V}_P = \frac{d}{dt}(R\mathbf{e}_r + Z\hat{k}) = \dot{R}\mathbf{e}_r + R\dot{\mathbf{e}}_r + \dot{Z}\hat{k} = \dot{R}\mathbf{e}_r + R\dot{\theta}\mathbf{e}_t + \dot{Z}\hat{k} .$$

Similarly, the acceleration A_p, which is the time derivative of the velocity V_p, is given by:

$$\mathbf{A}_P = \frac{d}{dt}(\dot{R}\mathbf{e}_r + R\dot{\theta}\mathbf{e}_t + \dot{Z}\hat{k}) = (\ddot{R} - R\dot{\theta}^2)\mathbf{e}_r + (R\ddot{\theta} + 2\dot{R}\dot{\theta})\mathbf{e}_t + \ddot{Z}\hat{k}.$$

The term $-R\dot{\theta}^2\mathbf{e}_r$ acts toward the center of curvature of the path at that point on the path, is commonly called the centripetal acceleration. The term $2\dot{R}\dot{\theta}\mathbf{e}_t$ is called the Coriolis acceleration.

Constant Radius

If the trajectory of the particle is constrained to lie on a cylinder, then the radius R is constant and the velocity and acceleration vectors simplify. The velocity of V_p is the time derivative of the trajectory P(t),

$$\mathbf{V}_P = \frac{d}{dt}(R\mathbf{e}_r + Z\hat{k}) = R\dot{\theta}\mathbf{e}_t + \dot{Z}\hat{k}.$$

The acceleration vector becomes:

$$\mathbf{A}_P = \frac{d}{dt}(R\dot{\theta}\mathbf{e}_t + \dot{Z}\hat{k}) = -R\dot{\theta}^2\mathbf{e}_r + R\ddot{\theta}\mathbf{e}_t + \ddot{Z}\hat{k}.$$

Planar Circular Trajectories

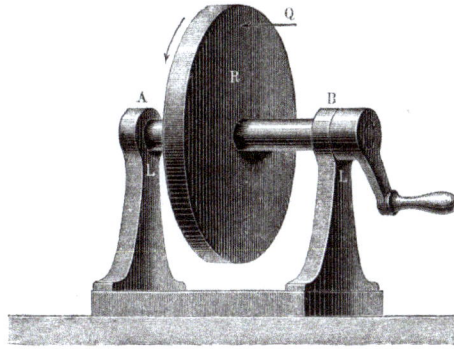

Each particle on the wheel travels in a planar circular trajectory (Kinematics of Machinery, 1876).

A special case of a particle trajectory on a circular cylinder occurs when there is no movement along the Z axis:

$$\mathbf{P}(t) = R\mathbf{e}_r + Z_0\hat{k},$$

where R and Z_0 are constants. In this case, the velocity V_P is given by:

$$\mathbf{V}_P = \frac{d}{dt}(R\mathbf{e}_r + Z_0\hat{k}) = R\dot{\theta}\mathbf{e}_t = R\omega\mathbf{e}_t,$$

where

$$\omega = \dot{\theta},$$

is the angular velocity of the unit vector e_t around the z axis of the cylinder.

The acceleration A_P of the particle P is now given by:

$$\mathbf{A}_P = \frac{d}{dt}(R\dot{\theta}\mathbf{e}_t) = -R\dot{\theta}^2\mathbf{e}_r + R\ddot{\theta}\mathbf{e}_t.$$

The components

$$a_r = -R\dot{\theta}^2, \quad a_t = R\ddot{\theta},$$

are called, respectively, the *radial* and *tangential components* of acceleration.

The notation for angular velocity and angular acceleration is often defined as

$$\omega = \dot{\theta}, \quad \alpha = \ddot{\theta},$$

so the radial and tangential acceleration components for circular trajectories are also written as

$$a_r = -R\omega^2, \quad a_t = R\alpha.$$

Point Trajectories in a Body Moving in The Plane

The movement of components of a mechanical system are analyzed by attaching a reference frame to each part and determining how the various reference frames move relative to each other. If the structural stiffness of the parts are sufficient, then their deformation can be neglected and rigid transformations can be used to define this relative movement. This reduces the description of the motion of the various parts of a complicated mechanical system to a problem of describing the geometry of each part and geometric association of each part relative to other parts.

Geometry is the study of the properties of figures that remain the same while the space is transformed in various ways---more technically, it is the study of invariants under a set of transformations. These transformations can cause the displacement of the triangle in the plane, while leaving the vertex angle and the distances between vertices unchanged. Kinematics is often described as applied geometry, where the movement of a mechanical system is described using the rigid transformations of Euclidean geometry.

The coordinates of points in a plane are two-dimensional vectors in R^2 (two dimensional space). Rigid transformations are those that preserve the distance between any two points. The set of rigid transformations in an n-dimensional space is called the special Euclidean group on R^n, and denoted *SE(n)*.

Displacements and Motion

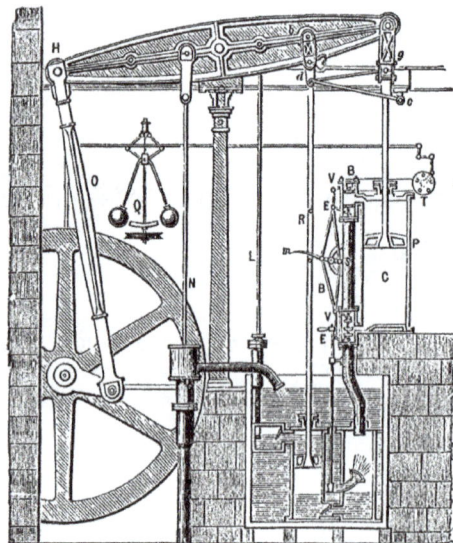

The movement of each of the components of the Boulton & Watt Steam Engine (1784) is modeled by a continuous set of rigid displacements.

The position of one component of a mechanical system relative to another is defined by introducing a reference frame, say *M*, on one that moves relative to a fixed frame, *F,* on the other. The rigid transformation, or displacement, of M relative to F defines the relative position of the two components. A displacement consists of the combination of a rotation and a translation.

The set of all displacements of M relative to F is called the configuration space of M. A smooth curve from one position to another in this configuration space is a continuous set of displacements, called the motion of M relative to F. The motion of a body consists of a continuous set of rotations and translations.

Matrix Representation

The combination of a rotation and translation in the plane R^2 can be represented by a certain type of 3x3 matrix known as a homogeneous transform. The 3x3 homogeneous transform is constructed from a 2x2 rotation matrix $A(\varphi)$ and the 2x1 translation vector $d=(d_x, d_y)$, as:

$$[T(\phi,\mathbf{d})] = \begin{bmatrix} A(\phi) & \mathbf{d} \\ 0 & 1 \end{bmatrix} = \begin{bmatrix} \cos\phi & -\sin\phi & d_x \\ \sin\phi & \cos\phi & d_y \\ 0 & 0 & 1 \end{bmatrix}.$$

These homogeneous transforms perform rigid transformations on the points in the plane z=1, that is on points with coordinates p=(x, y, 1).

In particular, let p define the coordinates of points in a reference frame M coincident with a fixed frame F. Then, when the origin of M is displaced by the translation vector d relative to the origin of F and rotated by the angle φ relative to the x-axis of F, the new coordinates in F of points in M are given by:

$$\mathbf{P} = [T(\phi,\mathbf{d})]\mathbf{p} = \begin{bmatrix} \cos\phi & -\sin\phi & d_x \\ \sin\phi & \cos\phi & d_y \\ 0 & 0 & 1 \end{bmatrix} \begin{Bmatrix} x \\ y \\ 1 \end{Bmatrix}.$$

Homogeneous transforms represent affine transformations. This formulation is necessary because a translation is not a linear transformation of R^2. However, using projective geometry, so that R^2 is considered to be a subset of R^3, translations become affine linear transformations.

Pure Translation

If a rigid body moves so that its reference frame M does not rotate ($\emptyset=0$) relative to the fixed frame F, the motion is said to be pure translation. In this case, the trajectory of every point in the body is an offset of the trajectory d(t) of the origin of M, that is:

$$\mathbf{P}(t) = [T(0,\mathbf{d}(t))]\mathbf{p} = \mathbf{d}(t) + \mathbf{p}.$$

Thus, for bodies in pure translation, the velocity and acceleration of every point P in the body are given by:

$$\mathbf{V}_P = \dot{\mathbf{P}}(t) = \dot{\mathbf{d}}(t) = \mathbf{V}_O, \quad \mathbf{A}_P = \ddot{\mathbf{P}}(t) = \ddot{\mathbf{d}}(t) = \mathbf{A}_O,$$

where the dot denotes the derivative with respect to time and V_O and A_O are the velocity and acceleration, respectively, of the origin of the moving frame M. Recall the coordinate vector p in M is constant, so its derivative is zero.

Rotation of a Body Around a Fixed Axis

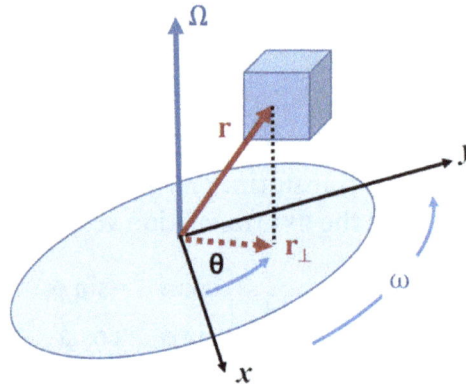

Figure 1: The angular velocity vector Ω points up for counterclockwise rotation and down for clockwise rotation, as specified by the right-hand rule. Angular position $\theta(t)$ changes with time at a rate $\omega(t) = d\theta/dt$.

Rotational or angular kinematics is the description of the rotation of an object. The description of rotation requires some method for describing orientation. Common descriptions include Euler angles and the kinematics of turns induced by algebraic products.

In what follows, attention is restricted to simple rotation about an axis of fixed orientation. The z-axis has been chosen for convenience.

Position

> This allows the description of a rotation as the angular position of a planar reference frame M relative to a fixed F about this shared z-axis. Coordinates p=(x, y) in M are related to coordinates P=(X, Y) in F by the matrix equation:

$$\mathbf{P}(t) = [A(t)]\mathbf{p},$$

> where

$$[A(t)] = \begin{bmatrix} \cos\theta(t) & -\sin\theta(t) \\ \sin\theta(t) & \cos\theta(t) \end{bmatrix},$$

> is the rotation matrix that defines the angular position of M relative to F as a function of time.

Velocity

> If the point p does not move in M, its velocity in F is given by

$$\mathbf{V}_P = \dot{\mathbf{P}} = [\dot{A}(t)]\mathbf{p}.$$

It is convenient to eliminate the coordinates p and write this as an operation on the trajectory P(t),

$$\mathbf{V}_P = [\dot{A}(t)][A(t)^{-1}]\mathbf{P} = [\Omega]\mathbf{P},$$

where the matrix

$$[\Omega] = \begin{bmatrix} 0 & -\omega \\ \omega & 0 \end{bmatrix},$$

is known as the angular velocity matrix of M relative to F. The parameter ω is the time derivative of the angle θ, that is:

$$\omega = \frac{d\theta}{dt}.$$

Acceleration

The acceleration of P(t) in F is obtained as the time derivative of the velocity,

$$\mathbf{A}_P = \ddot{P}(t) = [\dot{\Omega}]\mathbf{P} +, [\Omega]\dot{\mathbf{P}}$$

which becomes

$$\mathbf{A}_P = [\dot{\Omega}]\mathbf{P} + [\Omega][\Omega]\mathbf{P},$$

where

$$[\dot{\Omega}] = \begin{bmatrix} 0 & -\alpha \\ \alpha & 0 \end{bmatrix},$$

is the angular acceleration matrix of M on F, and

$$\alpha = \frac{d^2\theta}{dt^2}.$$

The description of rotation then involves these three quantities:

- Angular position : the oriented distance from a selected origin on the rotational axis to a point of an object is a vector r (t) locating the point. The vector r(t) has some projection (or, equivalently, some component) $r_\perp(t)$ on a plane perpendicular to the axis of rotation. Then the *angular position* of that point is the angle θ from a reference axis (typically the positive x-axis) to the vector $r_\perp(t)$ in a known rotation sense (typically given by the right-hand rule).

- Angular velocity : the angular velocity ω is the rate at which the angular position θ changes with respect to time t:

$$\omega = \frac{d\theta}{dt}$$

The angular velocity is represented in Figure 1 by a vector Ω pointing along the axis of rotation with magnitude ω and sense determined by the direction of rotation as given by the right-hand rule.

- Angular acceleration : the magnitude of the angular acceleration α is the rate at which the angular velocity ω changes with respect to time t:

$$\alpha = \frac{d\omega}{dt}$$

The equations of translational kinematics can easily be extended to planar rotational kinematics for constant angular acceleration with simple variable exchanges:

$$\omega_f = \omega_i + \alpha t$$

$$\theta_f - \theta_i = \omega_i t + \frac{1}{2}\alpha t^2$$

$$\theta_f - \theta_i = \tfrac{1}{2}(\omega_f + \omega_i)t$$

$$\omega_f^2 = \omega_i^2 + 2\alpha(\theta_f - \theta_i).$$

Here θ_i and θ_f are, respectively, the initial and final angular positions, ω_i and ω_f are, respectively, the initial and final angular velocities, and α is the constant angular acceleration. Although position in space and velocity in space are both true vectors (in terms of their properties under rotation), as is angular velocity, angle itself is not a true vector.

Point Trajectories in Body Moving in Three Dimensions

Important formulas in kinematics define the velocity and acceleration of points in a moving body as they trace trajectories in three-dimensional space. This is particularly important for the center of mass of a body, which is used to derive equations of motion using either Newton's second law or Lagrange's equations.

Position

In order to define these formulas, the movement of a component B of a mechanical system is defined by the set of rotations [A(t)] and translations d(t) assembled into the homogeneous transformation [T(t)]=[A(t), d(t)]. If p is the coordinates of a point P in B measured in the moving reference frame M, then the trajectory of this point traced in F is given by:

$$P(t) = [T(t)]\mathbf{p} = \begin{Bmatrix} \mathbf{P} \\ 1 \end{Bmatrix} = \begin{bmatrix} A(t) & \mathbf{d}(t) \\ 0 & 1 \end{bmatrix} \begin{Bmatrix} \mathbf{p} \\ 1 \end{Bmatrix}.$$

This notation does not distinguish between P = (X, Y, Z, 1), and P = (X, Y, Z), which is hopefully clear in context.

This equation for the trajectory of P can be inverted to compute the coordinate vector p in M as:

$$\mathbf{p} = [T(t)]^{-1}P(t) = \begin{Bmatrix} \mathbf{p} \\ 1 \end{Bmatrix} = \begin{bmatrix} A(t)^T & -A(t)^T\mathbf{d}(t) \\ 0 & 1 \end{bmatrix} \begin{Bmatrix} P(t) \\ 1 \end{Bmatrix}.$$

This expression uses the fact that the transpose of a rotation matrix is also its inverse, that is:

$$[A(t)]^T[A(t)] = I.$$

Velocity

The velocity of the point P along its trajectory $P(t)$ is obtained as the time derivative of this position vector,

$$\mathbf{V}_P = [\dot{T}(t)]\mathbf{p} = \begin{Bmatrix} \mathbf{V}_P \\ 0 \end{Bmatrix} = \begin{bmatrix} A(t) & d(t) \\ 0 & 1 \end{bmatrix}^{\cdot} \begin{Bmatrix} \mathbf{p} \\ 1 \end{Bmatrix} = \begin{bmatrix} \dot{A}(t) & \dot{d}(t) \\ 0 & 0 \end{bmatrix} \begin{Bmatrix} \mathbf{p} \\ 1 \end{Bmatrix}.$$

The dot denotes the derivative with respect to time; because p is constant, its derivative is zero.

This formula can be modified to obtain the velocity of P by operating on its trajectory $P(t)$ measured in the fixed frame F. Substituting the inverse transform for p into the velocity equation yields:

$$\mathbf{V}_P = [\dot{T}(t)][T(t)]^{-1}\mathbf{P}(t) = \begin{Bmatrix} \mathbf{V}_P \\ 0 \end{Bmatrix} = \begin{bmatrix} \dot{A} & \dot{d} \\ 0 & 0 \end{bmatrix} \begin{bmatrix} A & d \\ 0 & 1 \end{bmatrix}^{-1} \begin{Bmatrix} \mathbf{P}(t) \\ 1 \end{Bmatrix}$$

$$= \begin{bmatrix} \dot{A} & \dot{d} \\ 0 & 0 \end{bmatrix} A^{-1} \begin{bmatrix} 1 & -\mathbf{d} \\ 0 & A \end{bmatrix} \begin{Bmatrix} \mathbf{P}(t) \\ 1 \end{Bmatrix}$$

$$= \begin{bmatrix} \dot{A}A^{-1} & -\dot{A}A^{-1}\mathbf{d}+\dot{\mathbf{d}} \\ 0 & 0 \end{bmatrix} \begin{Bmatrix} \mathbf{P}(t) \\ 1 \end{Bmatrix}$$

$$= \begin{bmatrix} \dot{A}A^{T} & -\dot{A}A^{T}\mathbf{d}+\dot{\mathbf{d}} \\ 0 & 0 \end{bmatrix} \begin{Bmatrix} \mathbf{P}(t) \\ 1 \end{Bmatrix}$$

$$\mathbf{V}_P = [S]\mathbf{P}.$$

The matrix [S] is given by:

$$[S] = \begin{bmatrix} \Omega & -\Omega\mathbf{d}+\dot{\mathbf{d}} \\ 0 & 0 \end{bmatrix}$$

where

$$[\Omega] = \dot{A}A^{T},$$

is the angular velocity matrix.

Multiplying by the operator [S], the formula for the velocity V_P takes the form:

$$\mathbf{V}_P = [\Omega](\mathbf{P}-\mathbf{d})+\dot{\mathbf{d}} = \omega \times \mathbf{R}_{P/O} + \mathbf{V}_O,$$

where the vector ω is the angular velocity vector obtained from the components of the matrix $[\Omega]$; the vector

$$\mathbf{R}_{P/O} = \mathbf{P} - \mathbf{d},$$

is the position of P relative to the origin O of the moving frame M; and

$$\mathbf{V}_O = \dot{\mathbf{d}},$$

is the velocity of the origin O.

Acceleration

The acceleration of a point P in a moving body B is obtained as the time derivative of its velocity vector:

$$\mathbf{A}_P = \frac{d}{dt}\mathbf{V}_P = \frac{d}{dt}\left([S]\mathbf{P}\right) = [\dot{S}]\mathbf{P} + [S]\dot{\mathbf{P}} = [\dot{S}]\mathbf{P} + [S][S]\mathbf{P}.$$

This equation can be expanded firstly by computing

$$[\dot{S}] = \begin{bmatrix} \dot{\Omega} & -\dot{\Omega}\mathbf{d} - \Omega\dot{\mathbf{d}} + \ddot{\mathbf{d}} \\ 0 & 0 \end{bmatrix} = \begin{bmatrix} \dot{\Omega} & -\dot{\Omega}\mathbf{d} - \Omega\mathbf{V}_O + \mathbf{A}_O \\ 0 & 0 \end{bmatrix}$$

and

$$[S]^2 = \begin{bmatrix} \Omega & -\Omega\mathbf{d} + \mathbf{V}_O \\ 0 & 0 \end{bmatrix}^2 = \begin{bmatrix} \Omega^2 & -\Omega^2\mathbf{d} + \Omega\mathbf{V}_O \\ 0 & 0 \end{bmatrix}.$$

The formula for the acceleration $\mathbf{A}_P$ can now be obtained as:

$$\mathbf{A}_P = \dot{\Omega}(\mathbf{P} - \mathbf{d}) + \mathbf{A}_O + \Omega^2(\mathbf{P} - \mathbf{d}),$$

or

$$\mathbf{A}_P = \alpha \times \mathbf{R}_{P/O} + \omega \times \omega \times \mathbf{R}_{P/O} + \mathbf{A}_O,$$

where α is the angular acceleration vector obtained from the derivative of the angular velocity matrix;

$$\mathbf{R}_{P/O} = \mathbf{P} - \mathbf{d},$$

is the relative position vector (the position of P relative to the origin O of the moving frame M); and

$$\mathbf{A}_O = \ddot{\mathbf{d}}$$

is the acceleration of the origin of the moving frame M.

Kinematic Constraints

Kinematic constraints are constraints on the movement of components of a mechanical system. Kinematic constraints can be considered to have two basic forms, (i) constraints that arise from hinges, sliders and cam joints that define the construction of the system, called holonomic constraints, and (ii) constraints imposed on the velocity of the system such as the knife-edge constraint of ice-skates on a flat plane, or rolling without slipping of a disc or sphere in contact with a plane, which are called non-holonomic constraints. The following are some common examples.

Kinematic Coupling

A kinematic coupling exactly constrains all 6 degrees of freedom.

Rolling Without Slipping

An object that rolls against a surface without slipping obeys the condition that the velocity of its center of mass is equal to the cross product of its angular velocity with a vector from the point of contact to the center of mass:

$$\mathbf{v}_G(t) = \Omega \times \mathbf{r}_{G/O}.$$

For the case of an object that does not tip or turn, this reduces to $v = rw$

Inextensible Cord

This is the case where bodies are connected by an idealized cord that remains in tension and cannot change length. The constraint is that the sum of lengths of all segments of the cord is the total length, and accordingly the time derivative of this sum is zero. A dynamic problem of this type is the pendulum. Another example is a drum turned by the pull of gravity upon a falling weight attached to the rim by the inextensible cord. An *equilibrium* problem (i.e. not kinematic) of this type is the catenary.

Kinematic Pairs

Reuleaux called the ideal connections between components that form a machine kinematic pairs. He distinguished between higher pairs which were said to have line contact between the two links and lower pairs that have area contact between the links. J. Phillips shows that there are many ways to construct pairs that do not fit this simple classification.

Lower Pair

A lower pair is an ideal joint, or holonomic constraint, that maintains contact between a point, line or plane in a moving solid (three-dimensional) body to a corresponding point line or plane in the fixed solid body. There are the following cases:

A revolute pair, or hinged joint, requires a line, or axis, in the moving body to remain co-linear with a line in the fixed body, and a plane perpendicular to this line in the moving body maintain contact with a similar perpendicular plane in the fixed body. This imposes five constraints on the relative movement of the links, which therefore has one degree of freedom, which is pure rotation about the axis of the hinge.

A prismatic joint, or slider, requires that a line, or axis, in the moving body remain co-linear with a line in the fixed body, and a plane parallel to this line in the moving body maintain contact with a similar parallel plane in the fixed body. This imposes five constraints on the relative movement of the links, which therefore has one degree of freedom. This degree of freedom is the distance of the slide along the line.

A cylindrical joint requires that a line, or axis, in the moving body remain co-linear with a line in the fixed body. It is a combination of a revolute joint and a sliding joint. This joint has two degrees of freedom. The position of the moving body is defined by both the rotation about and slide along the axis.

A spherical joint, or ball joint, requires that a point in the moving body maintain contact with a point in the fixed body. This joint has three degrees of freedom.

A planar joint requires that a plane in the moving body maintain contact with a plane in fixed body. This joint has three degrees of freedom.

Higher Pairs

Generally speaking, a higher pair is a constraint that requires a curve or surface in the moving body to maintain contact with a curve or surface in the fixed body. For example, the contact between a cam and its follower is a higher pair called a *cam joint*. Similarly, the contact between the involute curves that form the meshing teeth of two gears are cam joints.

Kinematic Chains

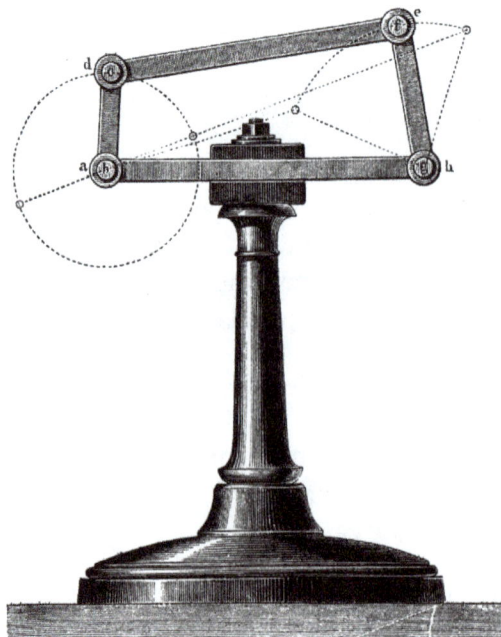

The_Kinematics_of_Machinery Kinematics of Machinery, 1876

Rigid bodies ("links") connected by kinematic pairs ("joints") are known as *kinematic chains*. Mechanisms and robots are examples of kinematic chains. The degree of freedom of a kinematic chain is computed from the number of links and the number and type of joints using the mobility formula. This formula can also be used to enumerate the topologies of kinematic chains that have a given degree of freedom, which is known as *type synthesis* in machine design.

Examples

The planar one degree-of-freedom linkages assembled from N links and j hinged or sliding joints are:

- $N=2, j=1$: a two-bar linkage that is the lever;

- $N=4, j=4$: the four-bar linkage;

- $N=6, j=7$: a six-bar linkage. This must have two links ("ternary links") that support three joints. There are two distinct topologies that depend on how the two ternary linkages are connected. In the Watt topology, the two ternary links have a common joint; in the Stephenson topology, the two ternary links do not have a common joint and are connected by binary links.

- $N=8, j=10$: eight-bar linkage with 16 different topologies;

- $N=10, j=13$: ten-bar linkage with 230 different topologies;

- $N=12, j=16$: twelve-bar linkage with 6,856 topologies.

Thermodynamics

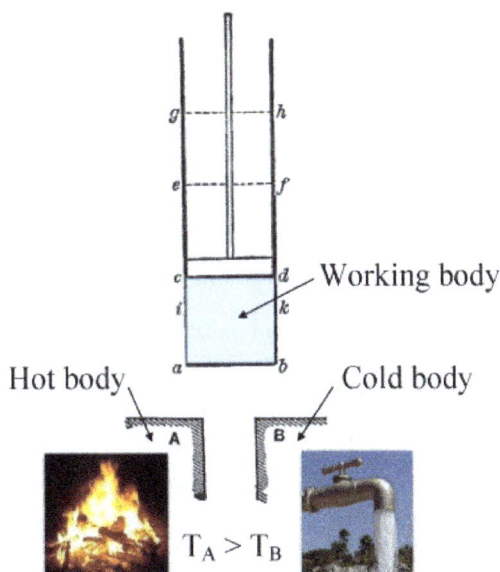

Annotated color version of the original 1824 Carnot heat engine showing the hot body (boiler), working body (system, steam), and cold body (water), the letters labeled according to the stopping points in Carnot cycle.

Thermodynamics is the branch of science concerned with heat and temperature and their relation to energy and work. It states that the behavior of these quantities is governed by the four laws of thermodynamics, irrespective of the composition or specific properties of the material or system in question. The laws of thermodynamics are explained in terms of microscopic constituents by statistical mechanics. Thermodynamics applies to a wide variety of topics in science and engineering, especially physical chemistry, chemical engineering and mechanical engineering.

Historically, thermodynamics developed out of a desire to increase the efficiency of early steam engines, particularly through the work of French physicist Nicolas Léonard Sadi Carnot (1824) who believed that engine efficiency was the key that could help France win the Napoleonic Wars. Scottish physicist Lord Kelvin was the first to formulate a concise definition of thermodynamics in 1854:

Thermo-dynamics is the subject of the relation of heat to forces acting between contiguous parts of bodies, and the relation of heat to electrical agency.

The initial application of thermodynamics to mechanical heat engines was extended early on to the study of chemical systems. Chemical thermodynamics studies the nature of the role of entropy in the process of chemical reactions and provided the bulk of expansion and knowledge of the field. Other formulations of thermodynamics emerged in the following decades. Statistical thermodynamics, or statistical mechanics, concerned itself with statistical predictions of the collective motion of particles from their microscopic behavior. In 1909, Constantin Carathéodory presented a purely mathematical approach to the field in his axiomatic formulation of thermodynamics, a description often referred to as *geometrical thermodynamics*.

Introduction

The starting point for most considerations of thermodynamic systems are the laws of thermodynamics, four principles that form an axiomatic basis. The first law specifies that energy can be exchanged between physical systems as heat and work. The second law defines the existence of the quantity entropy, a quantification of the state of order of a system that expresses the notion of useful work that can be performed.

In thermodynamics, interactions between large ensembles of objects are studied and categorized. Central to this are the concepts of *system* and *surroundings*. A system is composed of particles, whose average motions define its properties, which in turn are related to one another through equations of state. Properties can be combined to express internal energy and thermodynamic potentials, which are useful for determining conditions for equilibrium and spontaneous processes.

With these tools, thermodynamics can be used to describe how systems respond to changes in their environment. This can be applied to a wide variety of topics in science and engineering, such as engines, phase transitions, chemical reactions, transport phenomena, and even black holes. The results of thermodynamics are essential for other fields of physics and for chemistry, chemical engineering, aerospace engineering, mechanical engineering, cell biology, biomedical engineering, materials science, and economics, to name a few.

This article is focused mainly on classical thermodynamics which primarily studies systems in thermodynamic equilibrium. Non-equilibrium thermodynamics is often treated as an extension of the classical treatment, but statistical mechanics has brought many advances of the field.

École Polytechnique	Glasgow school	Berlin school	Edinburgh school
Sadi Carnot (1796-1832)	William Thomson (1824-1907)	Rudolf Clausius (1822-1888)	James Maxwell (1831-1879)

Vienna school	Gibbsian school	Dresden school	Dutch school
Ludwig Boltzmann (1844-1906)	Willard Gibbs (1839-1903)	Gustav Zeuner (1828-1907)	Johannes der Waals (1837-1923)

The thermodynamicists representative of the original eight founding schools of thermodynamics. The schools with the most-lasting effect in founding the modern versions of thermodynamics are the Berlin school, particularly as established in Rudolf Clausius's 1865 textbook *The Mechanical Theory of Heat*, the Vienna school, with the statistical mechanics of Ludwig Boltzmann, and the Gibbsian school at Yale University, American engineer Willard Gibbs' 1876 *On the Equilibrium of Heterogeneous Substances* launching chemical thermodynamics.

History

The history of thermodynamics as a scientific discipline generally begins with Otto von Guericke who, in 1650, built and designed the world's first vacuum pump and demonstrated a vacuum using his Magdeburg hemispheres. Guericke was driven to make a vacuum in order to disprove Aristotle's long-held supposition that 'nature abhors a vacuum'. Shortly after Guericke, the English physicist and chemist Robert Boyle had learned of Guericke's designs and, in 1656, in coordination with English scientist Robert Hooke, built an air pump. Using this pump, Boyle and Hooke noticed a correlation between pressure, temperature, and volume. In time, Boyle's Law was formulated, which states that pressure and volume are inversely proportional. Then, in 1679, based on these concepts, an associate of Boyle's named Denis Papin built a steam digester, which was a closed vessel with a tightly fitting lid that confined steam until a high pressure was generated.

Later designs implemented a steam release valve that kept the machine from exploding. By watching the valve rhythmically move up and down, Papin conceived of the idea of a piston and a cylinder engine. He did not, however, follow through with his design. Nevertheless, in 1697, based on Papin's designs, engineer Thomas Savery built the first engine, followed by Thomas Newcomen in 1712. Although these early engines were crude and inefficient, they attracted the attention of the leading scientists of the time.

The fundamental concepts of heat capacity and latent heat, which were necessary for the development of thermodynamics, were developed by Professor Joseph Black at the University of Glasgow, where James Watt was employed as an instrument maker. Black and Watt performed experiments together, but it was Watt who conceived the idea of the external condenser which resulted in a large increase in steam engine efficiency. Drawing on all the previous work led Sadi Carnot, the "father of thermodynamics", to publish *Reflections on the Motive Power of Fire* (1824), a discourse on heat, power, energy and engine efficiency. The paper outlined the basic energetic relations between the Carnot engine, the Carnot cycle, and motive power. It marked the start of thermodynamics as a modern science.

The first thermodynamic textbook was written in 1859 by William Rankine, originally trained as a physicist and a civil and mechanical engineering professor at the University of Glasgow. The first and second laws of thermodynamics emerged simultaneously in the 1850s, primarily out of the works of William Rankine, Rudolf Clausius, and William Thomson (Lord Kelvin).

The foundations of statistical thermodynamics were set out by physicists such as James Clerk Maxwell, Ludwig Boltzmann, Max Planck, Rudolf Clausius and J. Willard Gibbs.

During the years 1873-76 the American mathematical physicist Josiah Willard Gibbs published a series of three papers, the most famous being *On the Equilibrium of Heterogeneous Substances*, in which he showed how thermodynamic processes, including chemical reactions, could be graphically analyzed, by studying the energy, entropy, volume, temperature and pressure of the thermodynamic system in such a manner, one can determine if a process would occur spontaneously. Also Pierre Duhem in the 19th century wrote about chemical thermodynamics. During the early 20th century, chemists such as Gilbert N. Lewis, Merle Randall, and E. A. Guggenheim applied the mathematical methods of Gibbs to the analysis of chemical processes.

Etymology

The etymology of *thermodynamics* has an intricate history. It was first spelled in a hyphenated form as an adjective (*thermo-dynamic*) and from 1854 to 1868 as the noun *thermo-dynamics* to represent the science of generalized heat engines.

Pierre Perrot claims that the term *thermodynamics* was coined by James Joule in 1858 to designate the science of relations between heat and power., however, Joule never used that term, but used instead the term *perfect thermo-dynamic engine* in reference to Thomson's 1849 phraseology.

By 1858, *thermo-dynamics*, as a functional term, was used in William Thomson's paper *An Account of Carnot's Theory of the Motive Power of Heat*.

Branches of Description

The study of thermodynamical systems has developed into several related branches, each using a different fundamental model as a theoretical or experimental basis, or applying the principles to varying types of systems.

Classical Thermodynamics

Classical thermodynamics is the description of the states of thermodynamical systems at near-equilibrium, using macroscopic, empirical properties directly measurable in the laboratory. It is used to model exchanges of energy, work and heat based on the laws of thermodynamics. The qualifier *classical* reflects the fact that it represents the descriptive level in terms of macroscopic empirical parameters that can be measured in the laboratory, that was the first level of understanding in the 19th century. A microscopic interpretation of these concepts was provided by the development of statistical mechanics.

Statistical Mechanics

Statistical mechanics, also called statistical thermodynamics, emerged with the development of atomic and molecular theories in the late 19th century and early 20th century, supplementing thermodynamics with an interpretation of the microscopic interactions between individual particles or quantum-mechanical states. This field relates the microscopic properties of individual atoms and molecules to the macroscopic, bulk properties of materials that can be observed on the human scale, thereby explaining thermodynamics as a natural result of statistics, classical mechanics, and quantum theory at the microscopic level.

Chemical Thermodynamics

Chemical thermodynamics is the study of the interrelation of energy with chemical reactions or with a physical change of state within the confines of the laws of thermodynamics.

Treatment of Equilibrium

Equilibrium thermodynamics is the systematic study of transformations of matter and energy in systems as they approach equilibrium. The word equilibrium implies a state of balance. In an equilibrium state there are no unbalanced potentials, or driving forces, within the system. A central aim in equilibrium thermodynamics is: given a system in a well-defined initial state, subject to accurately specified constraints, to calculate what the state of the system will be once it has reached equilibrium.

Non-equilibrium thermodynamics is a branch of thermodynamics that deals with systems that are not in thermodynamic equilibrium. Most systems found in nature are not in thermodynamic equilibrium because they are not in stationary states, and are continuously and discontinuously subject to flux of matter and energy to and from other systems. The thermodynamic study of non-equilibrium systems requires more general concepts than are dealt with by equilibrium thermodynamics. Many natural systems still today remain beyond the scope of currently known macroscopic thermodynamic methods.

Laws of Thermodynamics

Thermodynamics is principally based on a set of four laws which are universally valid when applied to systems that fall within the constraints implied by each. In the various theoretical descriptions of thermodynamics these laws may be expressed in seemingly differing forms, but the most prominent formulations are the following:

- Zeroth law of thermodynamics: *If two systems are each in thermal equilibrium with a third, they are also in thermal equilibrium with each other.*

This statement implies that thermal equilibrium is an equivalence relation on the set of thermo-dynamic systems under consideration. Systems are said to be in equilibrium if the small, random exchanges between them (e.g. Brownian motion) do not lead to a net change in energy. This law is tacitly assumed in every measurement of temperature. Thus, if one seeks to decide if two bodies are at the same temperature, it is not necessary to bring them into contact and measure any chang-es of their observable properties in time. The law provides an empirical definition of temperature and justification for the construction of practical thermometers.

The zeroth law was not initially recognized as a law, as its basis in thermodynamical equilibrium was implied in the other laws. The first, second, and third laws had been explicitly stated prior and found common acceptance in the physics community. Once the importance of the zeroth law for the definition of temperature was realized, it was impracticable to renumber the other laws, hence it was numbered the *zeroth law*.

- First law of thermodynamics: *The internal energy of an isolated system is constant.*

The first law of thermodynamics is an expression of the principle of conservation of energy. It states that energy can be transformed (changed from one form to another), but cannot be created or destroyed.

The first law is usually formulated by saying that the change in the internal energy of a closed thermodynamic system is equal to the difference between the heat supplied to the system and the amount of work done by the system on its surroundings. It is important to note that internal energy is a state of the system whereas heat and work modify the state of the system. In other words, a specific internal energy of a system may be achieved by any combination of heat and work; the manner by which a system achieves a specific internal energy is path inde-pendent.

- Second law of thermodynamics: *Heat cannot spontaneously flow from a colder location to a hotter location.*

The second law of thermodynamics is an expression of the universal principle of decay observable in nature. The second law is an observation of the fact that over time, differences in temperature, pressure, and chemical potential tend to even out in a physical system that is isolated from the outside world. Entropy is a measure of how much this process has progressed. The entropy of an isolated system which is not in equilibrium will tend to increase over time, approaching a maxi-mum value at equilibrium.

In classical thermodynamics, the second law is a basic postulate applicable to any system involv-ing heat energy transfer; in statistical thermodynamics, the second law is a consequence of the assumed randomness of molecular chaos. There are many versions of the second law, but they all have the same effect, which is to explain the phenomenon of irreversibility in nature.

- Third law of thermodynamics: *As a system approaches absolute zero, all processes cease and the entropy of the system approaches a minimum value.*

The third law of thermodynamics is a statistical law of nature regarding entropy and the impossi-bility of reaching absolute zero of temperature. This law provides an absolute reference point for the determination of entropy. The entropy determined relative to this point is the absolute entropy. Alternate definitions are, "the entropy of all systems and of all states of a system is smallest at absolute zero," or equivalently "it is impossible to reach the absolute zero of temperature by any finite number of processes".

Absolute zero, at which all activity would stop if it were possible to happen, is −273.15 °C (degrees Celsius), or −459.67 °F (degrees Fahrenheit) or 0 K (kelvin).

System Models

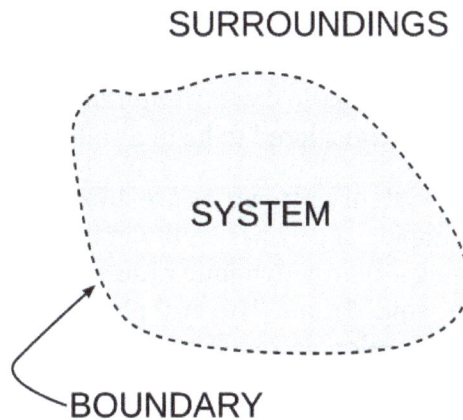

SURROUNDINGS

SYSTEM

BOUNDARY

A diagram of a generic thermodynamic system

An important concept in thermodynamics is the thermodynamic system, a precisely defined region of the universe under study. Everything in the universe except the system is known as the *surroundings*. A system is separated from the remainder of the universe by a *boundary* which may be notional or not, but which by convention delimits a finite volume. Exchanges of work, heat, or matter between the system and the surroundings take place across this boundary.

In practice, the boundary is simply an imaginary dotted line drawn around a volume when there is going to be a change in the internal energy of that volume. Anything that passes across the boundary that effects a change in the internal energy needs to be accounted for in the energy balance equation. The volume can be the region surrounding a single atom resonating energy, such as Max Planck defined in 1900; it can be a body of steam or air in a steam engine, such as Sadi Carnot defined in 1824; it can be the body of a tropical cyclone, such as Kerry Emanuel theorized in 1986 in the field of atmospheric thermodynamics; it could also be just one nuclide (i.e. a system of quarks) as hypothesized in quantum thermodynamics.

Boundaries are of four types: fixed, moveable, real, and imaginary. For example, in an engine, a fixed boundary means the piston is locked at its position; as such, a constant volume process occurs. In that same engine, a moveable boundary allows the piston to move in and out. For closed systems, boundaries are real while for open system boundaries are often imaginary.

Generally, thermodynamics distinguishes three classes of systems, defined in terms of what is allowed to cross their boundaries:

Interactions of thermodynamic systems			
Type of system	Mass flow	Work	Heat
Open	✓	✓	✓
Closed	✗	✓	✓
Thermally isolated	✗	✓	✗
Mechanically isolated	✗	✗	✓
Isolated	✗	✗	✗

As time passes in an isolated system, internal differences in the system tend to even out and pressures and temperatures tend to equalize, as do density differences. A system in which all equalizing processes have gone to completion is considered to be in a state of thermodynamic equilibrium.

In thermodynamic equilibrium, a system's properties are, by definition, unchanging in time. Systems in equilibrium are much simpler and easier to understand than systems which are not in equilibrium. Often, when analysing a thermodynamic process, it can be assumed that each intermediate state in the process is at equilibrium. This will also considerably simplify the situation. Thermodynamic processes which develop so slowly as to allow each intermediate step to be an equilibrium state are said to be reversible processes.

States and Processes

When a system is at equilibrium under a given set of conditions, it is said to be in a definite thermodynamic state. The state of the system can be described by a number of intensive variables and extensive variables. The properties of the system can be described by an equation of state which specifies the relationship between these variables. State may be thought of as the instantaneous quantitative description of a system with a set number of variables held constant.

A thermodynamic process may be defined as the energetic evolution of a thermodynamic system proceeding from an initial state to a final state. Typically, each thermodynamic process is distinguished from other processes in energetic character according to what parameters, such as temperature, pressure, or volume, etc., are held fixed. Furthermore, it is useful to group these processes into pairs, in which each variable held constant is one member of a conjugate pair.

Several commonly studied thermodynamic processes are:

- Isobaric process: occurs at constant pressure
- Isochoric process: occurs at constant volume (also called isometric/isovolumetric)
- Isothermal process: occurs at a constant temperature
- Adiabatic process: occurs without loss or gain of energy by heat
- Isentropic process: a reversible adiabatic process, occurs at a constant entropy
- Isenthalpic process: occurs at a constant enthalpy
- Steady state process: occurs without a change in the internal energy

Instrumentation

There are two types of thermodynamic instruments, the meter and the reservoir. A thermodynamic meter is any device which measures any parameter of a thermodynamic system. In some cases, the thermodynamic parameter is actually defined in terms of an idealized measuring instrument. For example, the zeroth law states that if two bodies are in thermal equilibrium with a third body, they are also in thermal equilibrium with each other. This principle, as noted by James Maxwell in 1872, asserts that it is possible to measure temperature. An idealized thermometer is a sample of an ideal gas at constant pressure. From the ideal gas law $pV=nRT$, the volume of such a sample can be used as an indicator of temperature; in this manner it defines temperature. Although pressure is defined mechanically, a pressure-measuring device, called a barometer may also be constructed from a sample of an ideal gas held at a constant temperature. A calorimeter is a device which is used to measure and define the internal energy of a system.

A thermodynamic reservoir is a system which is so large that it does not appreciably alter its state parameters when brought into contact with the test system. It is used to impose a particular value of a state parameter upon the system. For example, a pressure reservoir is a system at a particular pressure, which imposes that pressure upon any test system that it is mechanically connected to. The Earth's atmosphere is often used as a pressure reservoir.

Conjugate Variables

The central concept of thermodynamics is that of energy, the ability to do work. By the First Law, the total energy of a system and its surroundings is conserved. Energy may be transferred into a system by heating, compression, or addition of matter, and extracted from a system by cooling, expansion, or extraction of matter. In mechanics, for example, energy transfer equals the product of the force applied to a body and the resulting displacement.

Conjugate variables are pairs of thermodynamic concepts, with the first being akin to a "force" applied to some thermodynamic system, the second being akin to the resulting "displacement," and the product of the two equalling the amount of energy transferred. The common conjugate variables are:

- Pressure-volume (the mechanical parameters);
- Temperature-entropy (thermal parameters);
- Chemical potential-particle number (material parameters).

Potentials

Thermodynamic potentials are different quantitative measures of the stored energy in a system. Potentials are used to measure energy changes in systems as they evolve from an initial state to a final state. The potential used depends on the constraints of the system, such as constant temperature or pressure. For example, the Helmholtz and Gibbs energies are the energies available in a system to do useful work when the temperature and volume or the pressure and temperature are fixed, respectively.

The five most well known potentials are:

Name	Symbol	Formula	Natural variables
Internal energy	U	$\int (T\mathrm{d}S - p\mathrm{d}V + \sum_i \mu_i \mathrm{d}N_i)$	$S, V, \{N_i\}$
Helmholtz free energy	F	$U - TS$	$T, V, \{N_i\}$
Enthalpy	H	$U + pV$	$S, p, \{N_i\}$
Gibbs free energy	G	$U + pV - TS$	$T, p, \{N_i\}$
Landau Potential (Grand potential)	$\grave{U}, \Phi_G$	$U - TS - \sum_i \mu_i N_i$	$T, V, \{\mu_i\}$

where T is the temperature, S the entropy, p the pressure, V the volume, μ the chemical potential, N the number of particles in the system, and i is the count of particles types in the system.

Thermodynamic potentials can be derived from the energy balance equation applied to a thermodynamic system. Other thermodynamic potentials can also be obtained through Legendre transformation.

Structural Analysis

Structural analysis is the determination of the effects of loads on physical structures and their components.

Structures subject to this type of analysis include all that must withstand loads, such as buildings, bridges, vehicles, machinery, furniture, attire, soil strata, prostheses and biological tissue. Structural analysis employs the fields of applied mechanics, materials science and applied mathematics to compute a structure's deformations, internal forces, stresses, support reactions, accelerations, and stability. The results of the analysis are used to verify a structure's fitness for use, often precluding physical tests. Structural analysis is thus a key part of the engineering design of structures.

Structures and Loads

A structure refers to a body or system of connected parts used to support a load. Important examples related to Civil Engineering include buildings, bridges, and towers; and in other branches of engineering, ship and aircraft frames, tanks, pressure vessels, mechanical systems, and electrical supporting structures are important. In order to design a structure, one must serve a specified function for public use, the engineer must account for its safety, aesthetics, and serviceability, while taking into consideration economic and environmental constraints. Other branches of engineering work on a wide variety of non-building structures.

Classification of Structures

A *structural system* is the combination of structural elements and their materials. It is important for a structural engineer to be able to classify a structure by either its form or its function, by recognizing the various elements composing that structure. The structural elements guiding the systemic forces through the materials are not only such as a connecting rod, a truss, a beam, or a column, but also a cable, an arch, a cavity or channel, and even an angle, a surface structure, or a frame.

Loads

Once the dimensional requirement for a structure have been defined, it becomes necessary to determine the loads the structure must support. In order to design a structure, it is therefore necessary to first specify the loads that act on it. The design loading for a structure is often specified in building codes. There are two types of codes: general building codes and design codes, engineer must satisfy all the codes requirements for a reliable structure.

There are two types of loads that structure engineering must encounter in the design. First type of load is called Dead loads that consist of the weights of the various structural members and the weights of any objects that are permanently attached to the structure. For example, columns, beams, girders, the floor slab, roofing, walls, windows, plumbing, electrical fixtures, and other miscellaneous attachments. Second type of load is Live Loads which vary in their magnitude and location. There are many different types of live loads like building loads, highway bridge Loads, railroad bridge Loads, impact loads, wind loads, snow loads, earthquake loads, and other natural loads.

Analytical Methods

To perform an accurate analysis a structural engineer must determine such information as structural loads, geometry, support conditions, and materials properties. The results of such an analysis typically include support reactions, stresses and displacements. This information is then compared to criteria that indicate the conditions of failure. Advanced structural analysis may examine dynamic response, stability and non-linear behavior. There are three approaches to the analysis: the mechanics of materials approach (also known as strength of materials), the elasticity theory approach (which is actually a special case of the more general field of continuum mechanics), and the finite element approach. The first two make use of analytical formulations which apply mostly to simple linear elastic models, lead to closed-form solutions, and can often be solved by hand. The finite element approach is actually a numerical method for solving differential equations generated by theories of mechanics such as elasticity theory and strength of materials. However, the finite-element method depends heavily on the processing power of computers and is more applicable to structures of arbitrary size and complexity.

Regardless of approach, the formulation is based on the same three fundamental relations: equilibrium, constitutive, and compatibility. The solutions are approximate when any of these relations are only approximately satisfied, or only an approximation of reality.

Limitations

Each method has noteworthy limitations. The method of mechanics of materials is limited to very simple structural elements under relatively simple loading conditions. The structural elements and loading conditions allowed, however, are sufficient to solve many useful engineering problems. The theory of elasticity allows the solution of structural elements of general geometry under general loading conditions, in principle. Analytical solution, however, is limited to relatively simple cases. The solution of elasticity problems also requires the solution of a system of partial differential equations, which is considerably more mathematically demanding than the solution of mechanics of materials problems, which require at most the solution of an ordinary differential equation. The finite element method is perhaps the most restrictive and most useful at the same time. This method itself relies upon other structural theories (such as the other two discussed here) for equations to solve. It does, however, make it generally possible to solve these equations, even with highly complex geometry and loading conditions, with the restriction that there is always some numerical error. Effective and reliable use of this method requires a solid understanding of its limitations.

Strength of Materials Methods (Classical Methods)

The simplest of the three methods here discussed, the mechanics of materials method is available for simple structural members subject to specific loadings such as axially loaded bars, prismatic beams in a state of pure bending, and circular shafts subject to torsion. The solutions can under certain conditions be superimposed using the superposition principle to analyze a member undergoing combined loading. Solutions for special cases exist for common structures such as thin-walled pressure vessels.

For the analysis of entire systems, this approach can be used in conjunction with statics, giving rise to the *method of sections* and *method of joints* for truss analysis, moment distribution method for small rigid frames, and *portal frame* and *cantilever method* for large rigid frames. Except for moment distribution, which came into use in the 1930s, these methods were developed in their current forms in the second half of the nineteenth century. They are still used for small structures and for preliminary design of large structures.

The solutions are based on linear isotropic infinitesimal elasticity and Euler–Bernoulli beam theory. In other words, they contain the assumptions (among others) that the materials in question are elastic, that stress is related linearly to strain, that the material (but not the structure) behaves identically regardless of direction of the applied load, that all deformations are small, and that beams are long relative to their depth. As with any simplifying assumption in engineering, the more the model strays from reality, the less useful (and more dangerous) the result.

Example

There are 2 commonly used methods to find the truss element forces, namely the Method of Joints and the Method of Sections. Below is an example that is solved using both of these methods. The first diagram below is the presented problem for which we need to find the truss element forces. The second diagram is the loading diagram and contains the reaction forces from the joints.

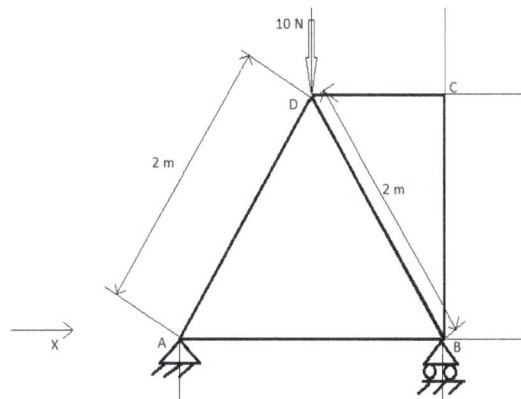

Since there is a pin joint at A, it will have 2 reaction forces. One in the x direction and the other in the y direction. At point B, we have a roller joint and hence we only have 1 reaction force in the y direction. Let us assume these forces to be in their respective positive directions (if they are not in the positive directions like we have assumed, then we will get a negative value for them).

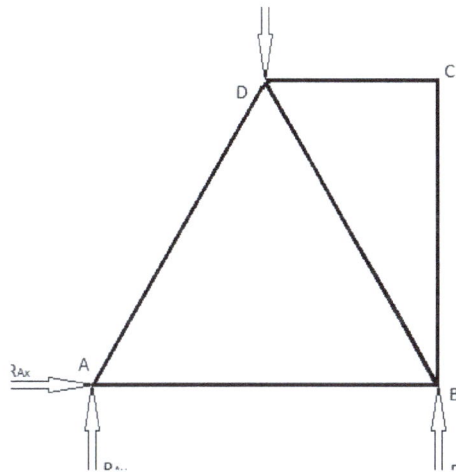

Since the system is in static equilibrium, the sum of forces in any direction is zero and the sum of moments about any point is zero. Therefore, the magnitude and direction of the reaction forces can be calculated.

$$\sum M_A = 0 = -10*1+2*R_B \Rightarrow R_B = 5$$

$$\sum F_y = 0 = R_{Ay} + R_B - 10 \Rightarrow R_{Ay} = 5$$

$$\sum F_x = 0 = R_{Ax}$$

Method of Joints

This method uses the force balance in the x and y directions at each of the joints in the truss structure.

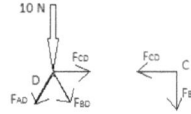

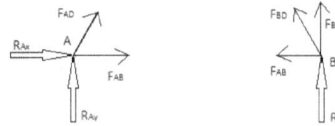

At A,

$$\sum F_y = 0 = R_{Ay} + F_{AD}\sin(60) = 5 + F_{AD}\frac{\sqrt{3}}{2} \Rightarrow F_{AD} = -\frac{10}{\sqrt{3}}$$

$$\sum F_x = 0 = R_{Ax} + F_{AD}\cos(60) + F_{AB} = 0 - \frac{10}{\sqrt{3}}\frac{1}{2} + F_{AB} \Rightarrow F_{AB} = \frac{5}{\sqrt{3}}$$

At D,

$$\sum F_y = 0 = -10 - F_{AD}\sin(60) - F_{BD}\sin(60) = -10 - \left(-\frac{10}{\sqrt{3}}\right)\frac{\sqrt{3}}{2} - F_{BD}\frac{\sqrt{3}}{2} \Rightarrow F_{BD} = -\frac{10}{\sqrt{3}}$$

$$\sum F_x = 0 = -F_{AD}\cos(60) + F_{BD}\cos(60) + F_{CD} = -\frac{10}{\sqrt{3}}\frac{1}{2} + \frac{10}{\sqrt{3}}\frac{1}{2} + F_{CD} \Rightarrow F_{CD} = 0$$

At C,

$$\sum F_y = 0 = -F_{BC} \Rightarrow F_{BC} = 0$$

Although we have found the forces in each of the truss elements, it is a good practice to verify the results by completing the remaining force balances.

$$\sum F_x = -F_{CD} = -0 = 0 \Rightarrow verified$$

At B,

$$\sum F_y = R_B + F_{BD}\sin(60) + F_{BC} = 5 + \left(-\frac{10}{\sqrt{3}}\right)\frac{\sqrt{3}}{2} + 0 = 0 \Rightarrow verified$$

$$\sum F_x = -F_{AB} - F_{BD}\cos(60) = \frac{5}{\sqrt{3}} - \frac{10}{\sqrt{3}}\frac{1}{2} = 0 \Rightarrow verified$$

Method of Sections

This method can be used when the truss element forces of only a few members are to be found. This method is used by introducing a single straight line cutting through the member whose force wants to be calculated. However this method has a limit in that the cutting line can pass through a maximum of only 3 members of the truss structure. This restriction is because this method uses the force balances in the x and y direction and the moment balance, which gives us a maximum of 3 equations to find a maximum of 3 unknown truss element forces through which this cut is made. Let us try to find the forces FAB, FBD and FCD in the above example

Method 1: Ignore The Right Side

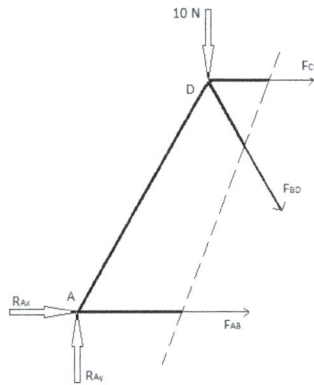

$$\sum M_D = 0 = -5*1 + \sqrt{3}*F_{AB} \Rightarrow F_{AB} = \frac{5}{\sqrt{3}}$$

$$\sum F_y = 0 = R_{Ay} - F_{BD}\sin(60) - 10 = 5 - F_{BD}\frac{\sqrt{3}}{2} - 10 \Rightarrow F_{BD} = -\frac{10}{\sqrt{3}}$$

$$\sum F_x = 0 = F_{AB} + F_{BD}\cos(60) + F_{CD} = \frac{5}{\sqrt{3}} - \frac{10}{\sqrt{3}}\frac{1}{2} + F_{CD} \Rightarrow F_{CD} = 0$$

Method 2: Ignore The Left Side

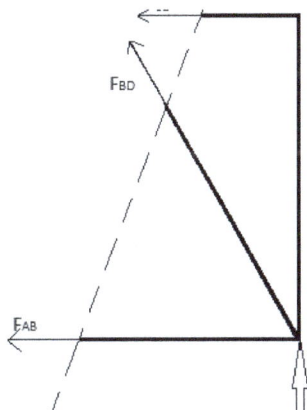

$$\sum M_B = 0 = \sqrt{3} * F_{CD} \Rightarrow F_{CD} = 0$$

$$\sum F_y = 0 = F_{BD}\sin(60) + R_B = F_{BD}\frac{\sqrt{3}}{2} + 5 \Rightarrow F_{BD} = -\frac{10}{\sqrt{3}}$$

$$\sum F_x = 0 = -F_{AB} - F_{BD}\cos(60) - F_{CD} = -F_{AB} - \left(-\frac{10}{\sqrt{3}}\right)\frac{1}{2} - 0 \Rightarrow F_{AB} = \frac{5}{\sqrt{3}}$$

The truss elements forces in the remaining members can be found by using the above method with a section passing through the remaining members.

Elasticity Methods

Elasticity methods are available generally for an elastic solid of any shape. Individual members such as beams, columns, shafts, plates and shells may be modeled. The solutions are derived from the equations of linear elasticity. The equations of elasticity are a system of 15 partial differential equations. Due to the nature of the mathematics involved, analytical solutions may only be produced for relatively simple geometries. For complex geometries, a numerical solution method such as the finite element method is necessary.

Methods Using Numerical Approximation

It is common practice to use approximate solutions of differential equations as the basis for structural analysis. This is usually done using numerical approximation techniques. The most commonly used numerical approximation in structural analysis is the Finite Element Method.

The finite element method approximates a structure as an assembly of elements or components with various forms of connection between them and each element of which has an associated stiffness. Thus, a continuous system such as a plate or shell is modeled as a discrete system with a finite number of elements interconnected at finite number of nodes and the overall stiffness is the result of the addition of the stiffness of the various elements. The behaviour of individual elements is characterized by the element's stiffness (or flexibility) relation. The assemblage of the various stiffness's into a master stiffness matrix that represents the entire structure leads to the system's stiffness or flexibility relation. To establish the stiffness (or flexibility) of a particular element, we can use the *mechanics of materials* approach for simple one-dimensional bar elements, and the *elasticity approach* for more complex two- and three-dimensional elements. The analytical and computational development are best effected throughout by means of matrix algebra, solving partial differential equations.

Early applications of matrix methods were applied to articulated frameworks with truss, beam and column elements; later and more advanced matrix methods, referred to as "finite element analysis", model an entire structure with one-, two-, and three-dimensional elements and can be used for articulated systems together with continuous systems such as a pressure vessel, plates, shells, and three-dimensional solids. Commercial computer software for structural analysis typically uses matrix finite-element analysis, which can be further classified into two main approaches: the displacement or stiffness method and the force or flexibility method. The stiffness method is the most popular by far thanks to its ease of implementation as well as of formulation for advanced applications. The finite-element technology is now sophisticated enough to handle just about any system as long as sufficient computing power is available. Its applicability includes, but is not limited to, linear and non-linear analysis, solid and fluid interactions, materials that are isotropic, orthotropic, or anisotropic, and external effects that are static, dynamic, and environmental factors. This, however, does not imply that the computed solution will automatically be reliable because much depends on the model and the reliability of the data input.

References

- Edmund Taylor Whittaker (1904). A Treatise on the Analytical Dynamics of Particles and Rigid Bodies. Cambridge University Press. Chapter 1. ISBN 0-521-35883-3.

- Russell C. Hibbeler (2009). "Kinematics and kinetics of a particle". Engineering Mechanics: Dynamics (12th ed.). Prentice Hall. p. 298. ISBN 0-13-607791-9.

- Ahmed A. Shabana (2003). "Reference kinematics". Dynamics of Multibody Systems (2nd ed.). Cambridge University Press. ISBN 978-0-521-54411-5.

- P. P. Teodorescu (2007). "Kinematics". Mechanical Systems, Classical Models: Particle Mechanics. Springer. p. 287. ISBN 1-4020-5441-6..

- Paul, Richard (1981). Robot manipulators: mathematics, programming, and control : the computer control of robot manipulators. MIT Press, Cambridge, MA. ISBN 978-0-262-16082-7.

- William Thomson Kelvin & Peter Guthrie Tait (1894). Elements of Natural Philosophy. Cambridge University Press. p. 4. ISBN 1-57392-984-0.

- M. Fogiel (1980). "Problem 17-11". The Mechanics Problem Solver. Research & Education Association. p. 613. ISBN 0-87891-519-2.

- Morris Kline (1990). Mathematical Thought from Ancient to Modern Times. Oxford University Press. p. 472. ISBN 0-19-506136-5.

- Phillips, Jack (2007). Freedom in Machinery, Volumes 1-2 (reprint ed.). Cambridge University Press. ISBN 978-0-521-67331-0.

- Tsai, Lung-Wen (2001). Mechanism design:enumeration of kinematic structures according to function (llustrated ed.). CRC Press. p. 121. ISBN 978-0-8493-0901-4.

- Clausius, Rudolf (1850). On the Motive Power of Heat, and on the Laws which can be deduced from it for the Theory of Heat. Poggendorff's Annalen der Physik, LXXIX (Dover Reprint). ISBN 0-486-59065-8.

- Van Ness, H.C. (1983) [1969]. Understanding Thermodynamics. Dover Publications, Inc. ISBN 9780486632773. OCLC 8846081.

- Smith, J.M.; Van Ness, H.C.; Abbott, M.M. (2005). Introduction to Chemical Engineering Thermodynamics. McGraw Hill. ISBN 0-07-310445-0. OCLC 56491111.

- Cengel, Yunus A.; Boles, Michael A. (2005). Thermodynamics - an Engineering Approach. McGraw-Hill. ISBN 0-07-310768-9.

- Gibbs, Willard (1993). The Scientific Papers of J. Willard Gibbs, Volume One: Thermodynamics. Ox Bow Press. ISBN 0-918024-77-3. OCLC 27974820.

- Perrot, Pierre (1998). A to Z of Thermodynamics. Oxford University Press. ISBN 0-19-856552-6. OCLC 123283342.

- Clark, John, O.E. (2004). The Essential Dictionary of Science. Barnes & Noble Books. ISBN 0-7607-4616-8. OCLC 58732844.

- Dugdale, J.S. (1998). Entropy and its Physical Meaning. Taylor and Francis. ISBN 0-7484-0569-0. OCLC 36457809.

- Haynie, Donald, T. (2001). Biological Thermodynamics. Cambridge University Press. ISBN 0-521-79549-4. OCLC 43993556.

- Chandramouli, P.N (2015). Structural Analysis I : Analysis of Statically Determinate Structures. Yes Dee Publishing Pvt Ltd. ISBN 9789380381473.

3

Principles of Mechanical Engineering

This chapter elucidates the principles of mechanical engineering, providing a complete understanding on the subject. The principles of mechanical engineering elucidated in this chapter are computer aided engineering, solid modeling, computer aided design and computational fluid dynamics. The section strategically encompasses all the principles of mechanical engineering, providing a complete understanding.

Computer-aided Engineering

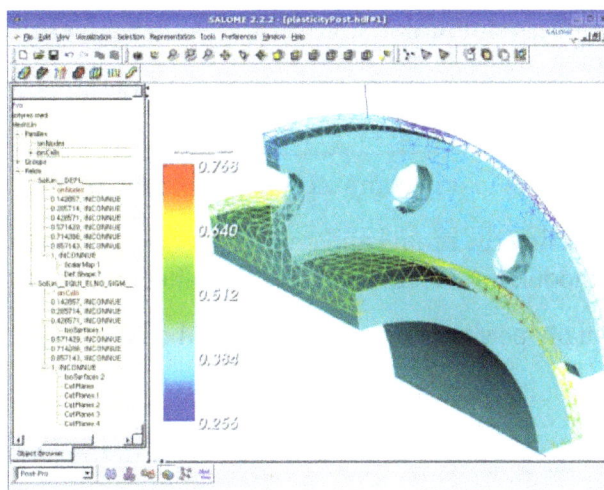

Nonlinear static analysis of a 3D structure subjected to plastic deformations

Computer-aided engineering (CAE) is the broad usage of computer software to aid in engineering analysis tasks. It includes finite element analysis (FEA), computational fluid dynamics (CFD), multibody dynamics (MBD), and optimization.

Overview

Software tools that have been developed to support these activities are considered CAE tools. CAE tools are being used, for example, to analyze the robustness and performance of components and assemblies. The term encompasses simulation, validation, and optimization of products and manufacturing tools. In the future, CAE systems will be major providers of information to help support design teams in decision making. Computer-aided engineering is used in many fields such as automotive, aviation, space, and shipbuilding industries.

In regard to information networks, CAE systems are individually considered a single node on a total information network and each node may interact with other nodes on the network.

CAE systems can provide support to businesses. This is achieved by the use of reference architectures and their ability to place information views on the business process. Reference architecture is the basis from which information model, especially product and manufacturing models.

The term CAE has also been used by some in the past to describe the use of computer technology within engineering in a broader sense than just engineering analysis. It was in this context that the term was coined by Jason Lemon, founder of SDRC in the late 1970s. This definition is however better known today by the terms CAx and PLM.

CAE Fields and Phases

- CAE areas covered include:

- Stress analysis on components and assemblies using Finite Element Analysis (FEA);

- Thermal and fluid flow analysis Computational fluid dynamics (CFD);

- Multibody dynamics (MBD) and Kinematics;

- Analysis tools for process simulation for operations such as casting, molding, and die press forming.

- Optimization of the product or process.

In general, there are three phases in any computer-aided engineering task:

- Pre-processing – defining the model and environmental factors to be applied to it. (typically a finite element model, but facet, voxel and thin sheet methods are also used)

- Analysis solver (usually performed on high powered computers)

- Post-processing of results (using visualization tools)

This cycle is iterated, often many times, either manually or with the use of commercial optimization software.

CAE in The Automotive Industry

CAE tools are very widely used in the automotive industry. In fact, their use has enabled the automakers to reduce product development cost and time while improving the safety, comfort, and durability of the vehicles they produce. The predictive capability of CAE tools has progressed to the point where much of the design verification is now done using computer simulations rather than physical prototype testing. CAE dependability is based upon all proper assumptions as inputs and must identify critical inputs (BJ). Even though there have been many advances in CAE, and it is widely used in the engineering field, physical testing is still a must. It is used for verification and model updating, to accurately define loads and boundary conditions and for final prototype sign-off.

The Future of CAE in The Product Development Process

Even though CAE has built a strong reputation as verification, troubleshooting and analysis tool, there is still a perception that sufficiently accurate results come rather late in the design cycle to really drive the design. This can expected to become a problem as modern products become ever more complex. They include smart systems, which leads to an increased need for multi-physics analysis including controls, and contain new lightweight materials, to which engineers are often less familiar. CAE software companies and manufacturers are constantly looking for tools and process improvements to change this situation. On the software side, they are constantly looking to develop more powerful solvers, better use computer resources and include engineering knowledge in pre- and post-processing. On the process side, they try to achieve a better alignment between 3D CAE, 1D System Simulation and physical testing. This should increase modeling realism and calculation speed. On top of that, they try to better integrate CAE in the overall product lifecycle management. In this way, they can connect product design with product use, which is an absolute must for smart products. Such an enhanced engineering process is also referred to as predictive engineering analytics.

Solid Modeling

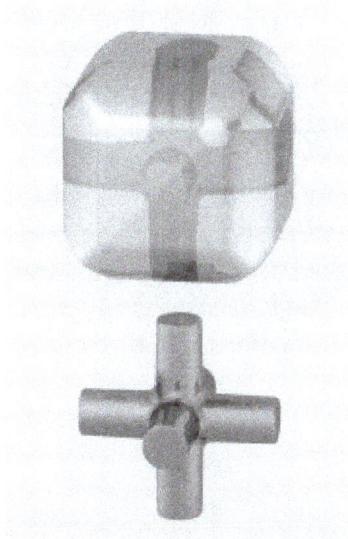

The geometry in solid modeling is fully described in 3D space; objects can be viewed from any angle.

Solid modeling (or modelling) is a consistent set of principles for mathematical and computer modeling of three-dimensional solids. Solid modeling is distinguished from related areas of geometric modeling and computer graphics by its emphasis on physical fidelity. Together, the principles of geometric and solid modeling form the foundation of computer-aided design and in general support the creation, exchange, visualization, animation, interrogation, and annotation of digital models of physical objects.

Overview

The use of solid modeling techniques allows for the automation of several difficult engineering calculations that are carried out as a part of the design process. Simulation, planning, and verification of processes such as machining and assembly were one of the main catalysts for the development of solid modeling. More recently, the range of supported manufacturing applications has been greatly expanded to include sheet metal manufacturing, injection molding, welding, pipe routing etc. Beyond traditional manufacturing, solid modeling techniques serve as the foundation for rapid prototyping, digital data archival and reverse engineering by reconstructing solids from sampled points on physical objects, mechanical analysis using finite elements, motion planning and NC path verification, kinematic and dynamic analysis of mechanisms, and so on. A central problem in all these applications is the ability to effectively represent and manipulate three-dimensional geometry in a fashion that is consistent with the physical behavior of real artifacts. Solid modeling research and development has effectively addressed many of these issues, and continues to be a central focus of computer-aided engineering.

Mathematical Foundations

The notion of solid modeling as practiced today relies on the specific need for informational completeness in mechanical geometric modeling systems, in the sense that any computer model should support all geometric queries that may be asked of its corresponding physical object. The requirement implicitly recognizes the possibility of several computer representations of the same physical object as long as any two such representations are consistent. It is impossible to computationally verify informational completeness of a representation unless the notion of a physical object is defined in terms of computable mathematical properties and independent of any particular representation. Such reasoning led to the development of the modeling paradigm that has shaped the field of solid modeling as we know it today.

All manufactured components have finite size and well behaved boundaries, so initially the focus was on mathematically modeling rigid parts made of homogeneous isotropic material that could be added or removed. These postulated properties can be translated into properties of subsets of three-dimensional Euclidean space. The two common approaches to define solidity rely on point-set topology and algebraic topology respectively. Both models specify how solids can be built from simple pieces or cells.

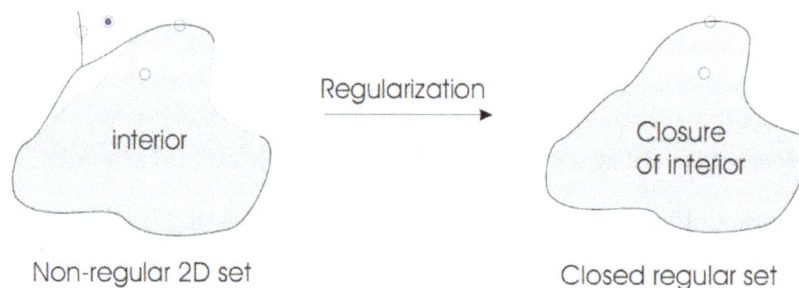

Regularization of a 2-d set by taking the closure of its interior

According to the continuum point-set model of solidity, all the points of any $X \subset R^3$ can be classified according to their neighborhoods with respect to X as interior, exterior, or boundary points. Assuming R^3 is endowed with the typical Euclidean metric, a neighborhood of a point $p \in X$ takes the form of an open ball. For X to be considered solid, every neighborhood of any $p \in X$ must be consistently three dimensional; points with lower-dimensional neighborhoods indicate a lack of solidity. Dimensional homogeneity of neighborhoods is guaranteed for the class of *closed regular* sets, defined as sets equal to the closure of their interior. Any $X \subset R^3$ can be turned into a closed regular set or *regularized* by taking the closure of its interior, and thus the modeling space of solids is mathematically defined to be the space of closed regular subsets of R^3 (by the Heine-Borel theorem it is implied that all solids are compact sets). In addition, solids are required to be closed under the Boolean operations of set union, intersection, and difference (to guarantee solidity after material addition and removal). Applying the standard Boolean operations to closed regular sets may not produce a closed regular set, but this problem can be solved by regularizing the result of applying the standard Boolean operations. The regularized set operations are denoted $\cup^*$, $\cap^*$, and $-^*$.

The combinatorial characterization of a set $X \subset R^3$ as a solid involves representing X as an orientable cell complex so that the cells provide finite spatial addresses for points in an otherwise innumerable continuum. The class of semi-analytic bounded subsets of Euclidean space is closed under Boolean operations (standard and regularized) and exhibits the additional property that every semi-analytic set can be stratified into a collection of disjoint cells of dimensions 0,1,2,3. A triangulation of a semi-analytic set into a collection of points, line segments, triangular faces, and tetrahedral elements is an example of a stratification that is commonly used. The combinatorial model of solidity is then summarized by saying that in addition to being semi-analytic bounded subsets, solids are three-dimensional topological polyhedra, specifically three-dimensional orientable manifolds with boundary. In particular this implies the Euler characteristic of the combinatorial boundary of the polyhedron is 2. The combinatorial manifold model of solidity also guarantees the boundary of a solid separates space into exactly two components as a consequence of the Jordan-Brouwer theorem, thus eliminating sets with non-manifold neighborhoods that are deemed impossible to manufacture.

The point-set and combinatorial models of solids are entirely consistent with each other, can be used interchangeably, relying on continuum or combinatorial properties as needed, and can be extended to n dimensions. The key property that facilitates this consistency is that the class of closed regular subsets of R^n coincides precisely with homogeneously n-dimensional topological polyhedra. Therefore, every n-dimensional solid may be unambiguously represented by its boundary and the boundary has the combinatorial structure of an $n-1$-dimensional polyhedron having homogeneously $n-1$-dimensional neighborhoods.

Solid Representation Schemes

Based on assumed mathematical properties, any scheme of representing solids is a method for capturing information about the class of semi-analytic subsets of Euclidean space. This means all representations are different ways of organizing the same geometric and topological data in the form of a data structure. All representation schemes are organized in terms of a finite number of operations on a set of primitives. Therefore, the modeling space of any particular representation is finite, and any single representation scheme may not completely suffice to represent all types of solids. For example, solids defined via combinations of regularized boolean operations cannot necessarily be represented as the sweep of a primitive moving according to a space trajectory, except in very simple cases. This forces modern geometric modeling systems to maintain several representation schemes of solids and also facilitate efficient conversion between representation schemes.

Below is a list of common techniques used to create or represent solid models. Modern modeling software may use a combination of these schemes to represent a solid.

Parameterized Primitive Instancing

This scheme is based on motion of families of objects, each member of a family distinguishable from the other by a few parameters. Each object family is called a *generic primitive*, and individual objects within a family are called *primitive instances*. For example, a family of bolts is a generic primitive, and a single bolt specified by a particular set of parameters is a primitive instance. The distinguishing characteristic of pure parameterized instancing schemes is the lack of means for combining instances to create new structures which represent new and more complex objects. The other main drawback of this scheme is the difficulty of writing algorithms for computing properties of represented solids. A considerable amount of family-specific information must be built into the algorithms and therefore each generic primitive must be treated as a special case, allowing no uniform overall treatment.

Spatial Occupancy Enumeration

This scheme is essentially a list of spatial *cells* occupied by the solid. The cells, also called voxels are cubes of a fixed size and are arranged in a fixed spatial grid (other polyhedral arrangements are also possible but cubes are the simplest). Each cell may be represented by the coordinates of a single point, such as the cell's centroid. Usually a specific scanning order is imposed and the corresponding ordered set of coordinates is called a *spatial array*. Spatial arrays are unambiguous and unique solid representations but are too verbose for use as 'master' or definitional representations. They can, however, represent coarse approximations of parts and can be used to improve the performance of geometric algorithms, especially when used in conjunction with other representations such as constructive solid geometry.

Cell Decomposition

This scheme follows from the combinatoric (algebraic topological) descriptions of solids detailed above. A solid can be represented by its decomposition into several cells. Spatial occupancy enumeration schemes are a particular case of cell decompositions where all the cells are cubical and lie in a regular grid. Cell decompositions provide convenient ways for computing certain topological properties of solids such as its connectedness (number of pieces) and genus (number of holes). Cell decompositions in the form of triangulations are the representations used in 3d finite elements for the numerical solution of partial differential equations. Other cell decompositions such as a Whitney regular stratification or Morse decompositions may be used for applications in robot motion planning.

Boundary Representation

In this scheme a solid is represented by the cellular decomposition of its boundary. Since the boundaries of solids have the distinguishing property that they separate space into regions defined by the interior of the solid and the complementary exterior according to the Jordan-Brouwer theorem discussed above, every point in space can unambiguously be tested against the solid by testing the point against the boundary of the solid. Recall that ability to test every point in the solid provides a guarantee of solidity. Using ray casting it is possible to count the number of intersections of a cast ray against the boundary of the solid. Even number of intersections correspond to exterior points, and odd number of intersections correspond to interior points. The assumption of boundaries as manifold cell complexes forces any boundary representation to obey disjointedness of distinct primitives, i.e. there are no self-intersections that cause non-manifold points. In particular, the manifoldness condition implies all pairs of vertices are disjoint, pairs of edges are either disjoint or intersect at one vertex, and pairs of faces are disjoint or intersect at a common edge. Several data structures that are combinatorial maps have been developed to store boundary representations of solids. In addition to planar faces, modern systems provide the ability to store quadrics and NURBS surfaces as a part of the boundary representation. Boundary representations have evolved into a ubiquitous representation scheme of solids in most commercial geometric modelers because of their flexibility in representing solids exhibiting a high level of geometric complexity.

Surface Mesh Modeling

Similar to boundary representation, the surface of the object is represented. However, rather than complex data structures and NURBS, a simple surface mesh of verticies and edges is used. Surface meshes can be structured (as in triangular meshes in STL files or quad meshes with horizontal and vertical rings of quadrilaterals), or unstructured meshes with randomly grouped triangles and higher level polygons.

Constructive Solid Geometry

Constructive solid geometry (CSG) connotes a family of schemes for representing rigid solids as Boolean constructions or combinations of primitives via the regularized set operations discussed above. CSG and boundary representations are currently the most important representation schemes for solids. CSG representations take the form of ordered binary trees where non-terminal nodes represent either rigid transformations (orientation preserving isometries) or regularized set operations. Terminal nodes are primitive leaves that represent closed regular sets. The semantics of CSG representations is clear. Each subtree represents a set resulting from applying the indicated transformations/regularized set operations on the set represented by the primitive leaves of the subtree. CSG representations are particularly useful for capturing design intent in the form of features corresponding to material addition or removal (bosses, holes, pockets etc.). The attractive properties of CSG include conciseness, guaranteed validity of solids, computationally convenient Boolean algebraic properties, and natural control of a solid's shape in terms of high level parameters defining the solid's primitives and their positions and orientations. The relatively simple data structure and elegant recursive algorithms have further contributed to the popularity of CSG.

Sweeping

The basic notion embodied in sweeping schemes is simple. A set moving through space may trace or *sweep* out volume (a solid) that may be represented by the moving set and its trajectory. Such a representation is important in the context of applications such as detecting the material removed from a cutter as it moves along a specified trajectory, computing dynamic interference of two solids undergoing relative motion, motion planning, and even in computer graphics applications such as tracing the motions of a brush moved on a canvas. Most commercial CAD systems provide (limited) functionality for constructing swept solids mostly in the form of a two dimensional cross section moving on a space trajectory transversal to the section. However, current research has shown several approximations of three dimensional shapes moving across one parameter, and even multi-parameter motions.

Implicit Representation

A very general method of defining a set of points X is to specify a predicate that can be evaluated at any point in space. In other words, X is defined *implicitly* to consist of all the points that satisfy the condition specified by the predicate. The simplest form of a predicate is the condition on the sign of a real valued function resulting in the familiar representation of sets by equalities and inequalities. For example, if $f = ax + by + cz + d$ the conditions $f(p) = 0$, $f(p) > 0$, and $f(p) < 0$ represent, respectively, a plane and two open linear halfspaces. More complex functional primitives may be defined by boolean combinations of simpler predicates. Furthermore, the theory of R-functions allow conversions of such representations into a single function inequality for any closed semi analytic set. Such a representation can be converted to a boundary representation using polygonization algorithms, for example, the marching cubes algorithm.

Parametric and Feature-Based Modeling

Features are defined to be parametric shapes associated with *attributes* such as intrinsic geometric parameters (length, width, depth etc.), position and orientation, geometric tolerances, material properties, and references to other features. Features also provide access to related production processes and resource models. Thus, features have a semantically higher level than primitive closed regular sets. Features are generally expected to form a basis for linking CAD with downstream manufacturing applications, and also for organizing databases for design data reuse. Parametric feature based modeling is frequently combined with constructive binary solid geometry (CSG) to fully describe systems of complex objects in engineering.

History of Solid Modelers

The historical development of solid modelers has to be seen in context of the whole history of CAD, the key milestones being the development of the research system BUILD followed by its commercial spin-off Romulus which went on to influence the development of Parasolid, ACIS and Solid Modeling Solutions. One of the first CAD developers in the Commonwealth of Independent States (CIS), ASCON began internal development of its own solid modeler in the 1990s. In November 2012, the mathematical division of ASCON became a separate company, and was named C3D Labs. It was assigned the task of developing the C3D geometric modeling kernel as a standalone product — the only commercial 3D modeling kernel from Russia. Other contributions came from Mäntylä, with his GWB and from the GPM project which contributed, among other things, hybrid modeling techniques at the beginning of the 1980s. This is also when the Programming Language of Solid Modeling PLaSM was conceived at the University of Rome.

Computer-Aided Design

The modeling of solids is only the minimum requirement of a CAD system's capabilities. Solid modelers have become commonplace in engineering departments in the last ten years due to faster computers and competitive software pricing. Solid modeling software creates a virtual 3D representation of components for machine design and analysis. A typical graphical user interface includes programmable macros, keyboard shortcuts and dynamic model manipulation. The ability to dynamically re-orient the model, in real-time shaded 3-D, is emphasized and helps the designer maintain a mental 3-D image.

A solid part model generally consists of a group of features, added one at a time, until the model is complete. Engineering solid models are built mostly with sketcher-based features; 2-D sketches that are swept along a path to become 3-D. These may be cuts, or extrusions for example. Design work on components is usually done within the context of the whole product using assembly modeling methods. An assembly model incorporates references to individual part models that comprise the product.

Another type of modeling technique is 'surfacing' (Freeform surface modeling). Here, surfaces are defined, trimmed and merged, and filled to make solid. The surfaces are usually defined with datum curves in space and a variety of complex commands. Surfacing is more difficult, but better applicable to some manufacturing techniques, like injection molding. Solid models for injection molded parts usually have both surfacing and sketcher based features.

Engineering drawings can be created semi-automatically and reference the solid models.

Parametric Modeling

Parametric modeling uses parameters to define a model (dimensions, for example). Examples of parameters are: dimensions used to create model features, material density, formulas to describe swept features, imported data (that describe a reference surface, for example). The parameter may be modified later, and the model will update to reflect the modification. Typically, there is a relationship between parts, assemblies, and drawings. A part consists of multiple features, and an assembly consists of multiple parts. Drawings can be made from either parts or assemblies.

Example: A shaft is created by extruding a circle 100 mm. A hub is assembled to the end of the shaft. Later, the shaft is modified to be 200 mm long (click on the shaft, select the length dimension, modify to 200). When the model is updated the shaft will be 200 mm long, the hub will relocate to the end of the shaft to which it was assembled, and the engineering drawings and mass properties will reflect all changes automatically.

Related to parameters, but slightly different are constraints. Constraints are relationships between entities that make up a particular shape. For a window, the sides might be defined as being parallel, and of the same length. Parametric modeling is obvious and intuitive. But for the first three decades of CAD this was not the case. Modification meant re-draw, or add a new cut or protrusion on top of old ones. Dimensions on engineering drawings were *created*, instead of *shown*. Parametric modeling is very powerful, but requires more skill in model creation. A complicated model for an injection molded part may have a thousand features, and modifying an early feature may cause later features to fail. Skillfully created parametric models are easier to maintain and modify. Parametric modeling also lends itself to data re-use. A whole family of capscrews can be contained in one model, for example.

Medical Solid Modeling

Modern computed axial tomography and magnetic resonance imaging scanners can be used to create solid models of internal body features, so-called volume rendering. Optical 3D scanners can be used to create point clouds or polygon mesh models of external body features.

Uses of medical solid modeling;

- Visualization

- Visualization of specific body tissues (just blood vessels and tumor, for example)

- Designing prosthetics, orthotics, and other medical and dental devices (this is sometimes called mass customization)

- Creating polygon mesh models for rapid prototyping (to aid surgeons preparing for difficult surgeries, for example)

- Combining polygon mesh models with CAD solid modeling (design of hip replacement parts, for example)

- Computational analysis of complex biological processes, e.g. air flow, blood flow

- Computational simulation of new medical devices and implants *in vivo*

If the use goes beyond visualization of the scan data, processes like image segmentation and image-based meshing will be necessary to generate an accurate and realistic geometrical description of the scan data.

Engineering

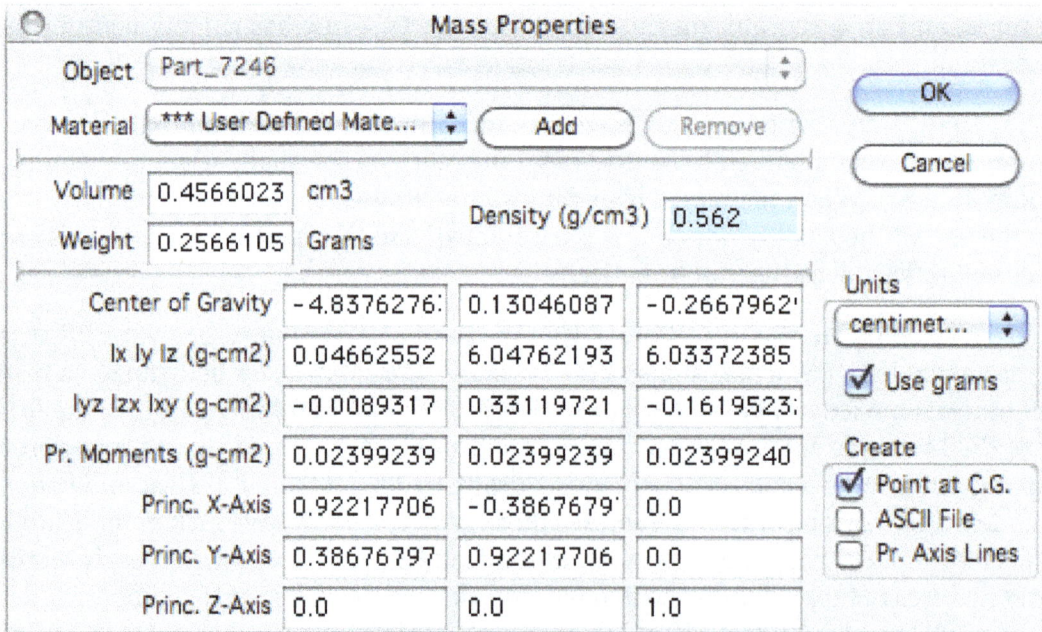

Mass properties window of a model in Cobalt

Because CAD programs running on computers "understand" the true geometry comprising complex shapes, many attributes of/for a 3D solid, such as its center of gravity, volume, and mass, can be quickly calculated. For instance, the cube shown at the top of this article measures 8.4 mm from flat to flat. Despite its many radii and the shallow pyramid on each of its six faces, its properties are readily calculated for the designer, as shown in the screenshot at right.

Computer-aided Design

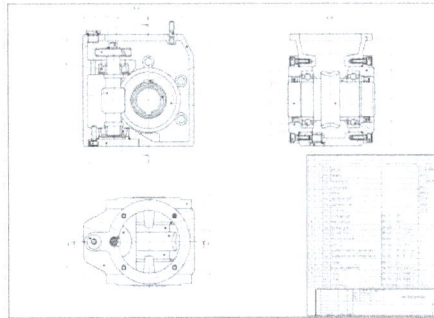
2D CAD drawing

Computer-aided drafting (CAD) is the use of computer systems to aid in the creation, modification, analysis, or optimization of a design. CAD software is used to increase the productivity of the designer, improve the quality of design, improve communications through documentation, and to create a database for manufacturing. CAD output is often in the form of electronic files for print, machining, or other manufacturing operations. The term CADD, (for *Computer Aided Design and Drafting*) is also used.

3D CAD model

Its use in designing electronic systems is known as electronic design automation, or EDA. In mechanical design it is known as mechanical design automation (MDA) or computer-aided drafting (CAD), which includes the process of creating a technical drawing with the use of computer software.

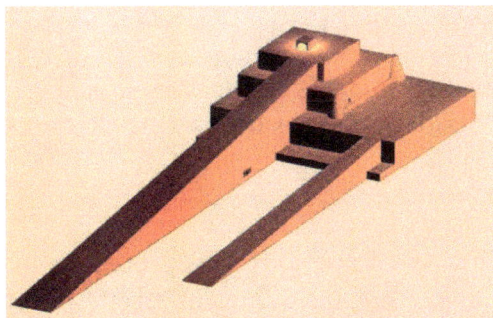
CAD rendering of Sialk ziggurat based on archeological evidence

CAD software for mechanical design uses either vector-based graphics to depict the objects of traditional drafting, or may also produce raster graphics showing the overall appearance of designed objects. However, it involves more than just shapes. As in the manual drafting of technical and engineering drawings, the output of CAD must convey information, such as materials, processes, dimensions, and tolerances, according to application-specific conventions.

CAD may be used to design curves and figures in two-dimensional (2D) space; or curves, surfaces, and solids in three-dimensional (3D) space.

CAD is an important industrial art extensively used in many applications, including automotive, shipbuilding, and aerospace industries, industrial and architectural design, prosthetics, and many more. CAD is also widely used to produce computer animation for special effects in movies, advertising and technical manuals, often called DCC digital content creation. The modern ubiquity and power of computers means that even perfume bottles and shampoo dispensers are designed using techniques unheard of by engineers of the 1960s. Because of its enormous economic importance, CAD has been a major driving force for research in computational geometry, computer graphics (both hardware and software), and discrete differential geometry.

The design of geometric models for object shapes, in particular, is occasionally called *computer-aided geometric design* (*CAGD*).

Overview of CAD Software

Starting around the mid 1970s, as computer aided design systems began to provide more capability than just an ability to reproduce manual drafting with electronic drafting, the cost benefit for companies to switch to CAD became apparent. The benefit of CAD systems over manual drafting are the capabilities one often takes for granted from computer systems today; automated generation of Bill of Material, auto layout in integrated circuits, interference checking, and many others. Eventually CAD provided the designer with the ability to perform engineering calculations. During this transition, calculations were still performed either by hand or by those individuals who could run computer programs. CAD was a revolutionary change in the engineering industry, where draftsmen, designers and engineering roles begin to merge. It did not eliminate departments, as much as it merged departments and empowered draftsman, designers and engineers. CAD is just another example of the pervasive effect computers were beginning to have on industry. Current computer-aided design software packages range from 2D vector-based drafting systems to 3D solid and surface modelers. Modern CAD packages can also frequently allow rotations in three dimensions, allowing viewing of a designed object from any desired angle, even from the inside looking out. Some CAD software is capable of dynamic mathematical modeling, in which case it may be marketed as CAD.

CAD technology is used in the design of tools and machinery and in the drafting and design of all types of buildings, from small residential types (houses) to the largest commercial and industrial structures (hospitals and factories).

CAD is mainly used for detailed engineering of 3D models and/or 2D drawings of physical components, but it is also used throughout the engineering process from conceptual design and layout of products, through strength and dynamic analysis of assemblies to definition of manufacturing methods of components. It can also be used to design objects. Furthermore, many CAD applications now offer advanced rendering and animation capabilities so engineers can better visualize their product designs. 4D BIM is a type of virtual construction engineering simulation incorporating time or schedule related information for project management.

CAD has become an especially important technology within the scope of computer-aided technologies, with benefits such as lower product development costs and a greatly shortened design cycle. CAD enables designers to layout and develop work on screen, print it out and save it for future editing, saving time on their drawings.

Uses

Computer-aided design is one of the many tools used by engineers and designers and is used in many ways depending on the profession of the user and the type of software in question.

CAD is one part of the whole Digital Product Development (DPD) activity within the Product Lifecycle Management (PLM) processes, and as such is used together with other tools, which are either integrated modules or stand-alone products, such as:

- Computer-aided engineering (CAE) and Finite element analysis (FEA)
- Computer-aided manufacturing (CAM) including instructions to Computer Numerical Control (CNC) machines
- Photorealistic rendering and Motion Simulation.
- Document management and revision control using Product Data Management (PDM).

CAD is also used for the accurate creation of photo simulations that are often required in the preparation of Environmental Impact Reports, in which computer-aided designs of intended buildings are superimposed into photographs of existing environments to represent what that locale will be like, where the proposed facilities are allowed to be built. Potential blockage of view corridors and shadow studies are also frequently analyzed through the use of CAD.

CAD has been proven to be useful to engineers as well. Using four properties which are history, features, parameterization, and high level constraints. The construction history can be used to look back into the model's personal features and work on the single area rather than the whole model. Parameters and constraints can be used to determine the size, shape, and other properties of the different modeling elements. The features in the CAD system can be used for the variety of tools for measurement such as tensile strength, yield strength, electrical or electro-magnetic properties. Also its stress, strain, timing or how the element gets affected in certain temperatures, etc.

Types

There are several different types of CAD, each requiring the operator to think differently about how to use them and design their virtual components in a different manner for each.

There are many producers of the lower-end 2D systems, including a number of free and open source programs. These provide an approach to the drawing process without all the fuss over scale and placement on the drawing sheet that accompanied hand drafting, since these can be adjusted as required during the creation of the final draft.

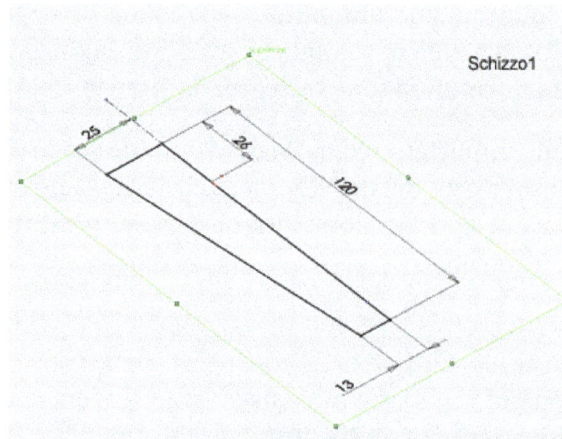

A simple procedure

3D wireframe is basically an extension of 2D drafting (not often used today). Each line has to be manually inserted into the drawing. The final product has no mass properties associated with it and cannot have features directly added to it, such as holes. The operator approaches these in a similar fashion to the 2D systems, although many 3D systems allow using the wireframe model to make the final engineering drawing views.

3D "dumb" solids are created in a way analogous to manipulations of real world objects (not often used today). Basic three-dimensional geometric forms (prisms, cylinders, spheres, and so on) have solid volumes added or subtracted from them, as if assembling or cutting real-world objects. Two-dimensional projected views can easily be generated from the models. Basic 3D solids don't usually include tools to easily allow motion of components, set limits to their motion, or identify interference between components.

There are two types of *3D Solid Modeling*

- *Parametric modeling* allows the operator to use what is referred to as "design intent". The objects and features created are modifiable. Any future modifications can be made by changing how the original part was created. If a feature was intended to be located from the center of the part, the operator should locate it from the center of the model. The feature could be located using any geometric object already available in the part, but this random placement would defeat the design intent. If the operator designs the part as it functions the parametric modeler is able to make changes to the part while maintaining geometric and functional relationships.

- *Direct or Explicit modeling* provide the ability to edit geometry without a history tree. With direct modeling once a sketch is used to create geometry the sketch is incorporated into the new geometry and the designer just modifies the geometry without needing the original sketch. As with parametric modeling, direct modeling has the ability to include relationships between selected geometry (e.g., tangency, concentricity).

Top end systems offer the capabilities to incorporate more organic, aesthetics and ergonomic features into designs. Freeform surface modeling is often combined with solids to allow the designer to create products that fit the human form and visual requirements as well as they interface with the machine.

Technology

A CAD model of a computer mouse.

Originally software for Computer-Aided Design systems was developed with computer languages such as Fortran, ALGOL but with the advancement of object-oriented programming methods this has radically changed. Typical modern parametric feature based modeler and freeform surface systems are built around a number of key C modules with their own APIs. A CAD system can be seen as built up from the interaction of a graphical user interface (GUI) with NURBS geometry and/or boundary representation (B-rep) data via a geometric modeling kernel. A geometry constraint engine may also be employed to manage the associative relationships between geometry, such as wireframe geometry in a sketch or components in an assembly.

Unexpected capabilities of these associative relationships have led to a new form of prototyping called digital prototyping. In contrast to physical prototypes, which entail manufacturing time in the design. That said, CAD models can be generated by a computer after the physical prototype has been scanned using an industrial CT scanning machine. Depending on the nature of the business, digital or physical prototypes can be initially chosen according to specific needs.

Today, CAD systems exist for all the major platforms (Windows, Linux, UNIX and Mac OS X); some packages even support multiple platforms.

Right now, no special hardware is required for most CAD software. However, some CAD systems can do graphically and computationally intensive tasks, so a modern graphics card, high speed (and possibly multiple) CPUs and large amounts of RAM may be recommended.

The human-machine interface is generally via a computer mouse but can also be via a pen and digitizing graphics tablet. Manipulation of the view of the model on the screen is also sometimes done with the use of a Spacemouse/SpaceBall. Some systems also support stereoscopic glasses for viewing the 3D model.Technologies which in the past were limited to larger installations or specialist applications have become available to a wide group of users.These include the CAVE or HMD`s and interactive devices like motion-sensing technology

Software

CAD software enables engineers and architects to design, inspect and manage engineering projects within an integrated graphical user interface (GUI) on a personal computer system. Most applications support solid modeling with boundary representation (B-Rep) and NURBS geometry, and enable the same to be published in a variety of formats. A geometric modeling kernel is a software component that provides solid modeling and surface modeling features to CAD applications.

Based on market statistics, commercial software from Autodesk, Dassault Systems, Siemens PLM Software and PTC dominate the CAD industry. The following is a list of major CAD applications, grouped by usage statistics.

Commercial

- Autodesk AutoCAD
- Autodesk Inventor
- Dassault CATIA
- Dassault SolidWorks
- Kubotek KeyCreator
- Siemens NX
- Siemens Solid Edge
- PTC Pro/ENGINEER (now renamed Creo)
- IronCAD
- MEDUSA
- ProgeCAD
- SpaceClaim
- Rhinoceros 3D
- VariCAD
- VectorWorks
- Cobalt

Free and open source

- 123D
- LibreCAD
- FreeCAD
- BRL-CAD
- OpenSCAD
- NanoCAD
- QCad

CAD Kernels

- Parasolid by Siemens

- ACIS by Spatial

- KCM by Kubotek

- ShapeManager by Autodesk

- Open CASCADE

- C3D by C3D Labs

History

Designers have long used computers for their calculations. Digital computers were used in power system analysis or optimization as early as proto-"Whirlwind" in 1949. Circuit design theory, or power network methodology would be algebraic, symbolic, and often vector-based. Examples of problems being solved in the mid-1940s to 50s include, Servo motors controlled by generated pulse (1949), The digital computer with built-in compute operations to automatically co-ordinate transforms to compute radar related vectors (1951) and the essentially graphic mathematical process of forming a shape with a digital machine tool (1952). These were accomplished with the use of computer software. The man credited with coining the term CAD. Douglas T. Ross stated "As soon as I saw the interactive display equipment, [being used by radar operators 1953]. The designers of these very early computers built utility programs so that programmers could debug programs using flow charts on a display scope with logical switches that could be opened and closed during the debugging session. They found that they could create electronic symbols and geometric figures to be used to create simple circuit diagrams and flow charts. They made the pleasant discovery that an object once drawn could be reproduced at will, its orientation, Linkage [flux, mechanical, lexical scoping] or scale changed. This suggested numerous possibilities to them. It took ten years of interdisciplinary development work before SKETCHPAD sitting on evolving math libraries emerged from MIT`s labs. Additional developments were carried out in the 1960s within the aircraft, automotive, industrial control and electronics industries in the area of 3D surface construction, NC programming and design analysis, most of it independent of one another and often not publicly published until much later. Some of the mathematical description work on curves was developed in the early 1940s by Robert Issac Newton from Pawtucket, Rhode Island. Robert A. Heinlein in his 1957 novel *The Door into Summer* suggested the possibility of a robotic *Drafting Dan*. However, probably the most important work on polynomial curves and sculptured surface was done by Pierre Bézier, Paul de Casteljau (Citroen), Steven Anson Coons (MIT, Ford), James Ferguson (Boeing), Carl de Boor (GM), Birkhoff (GM) and Garibedian (GM) in the 1960s and W. Gordon (GM) and R. Riesenfeld in the 1970s.

The invention of the 3D CAD/CAM is attributed to a French engineer, Pierre Bezier (Arts et Métiers ParisTech, Renault). After his mathematical work concerning surfaces, he developed UNISURF, between 1966 and 1968, to ease the design of parts and tools for the automotive industry. Then, UNISURF became the working base for the following generations of CAD software.

It is argued that a turning point was the development of the SKETCHPAD system at MIT by Ivan Sutherland (who later created a graphics technology company with Dr. David Evans). The distinctive feature of SKETCHPAD was that it allowed the designer to interact with his computer graphically: the design can be fed into the computer by drawing on a CRT monitor with a light pen. Effectively, it was a prototype of graphical user interface, an indispensable feature of modern CAD. Sutherland presented his paper Sketchpad: A Man-Machine Graphical Communication System in 1963 at a Joint Computer Conference having worked on it as his PhD thesis paper for a few years. Quoting,"For drawings where motion of the drawing, or analysis of a drawn problem is of value to the user, Sketchpad excels. For highly repetitive drawings or drawings where accuracy is required, Sketchpad is sufficiently faster than conventional techniques to be worthwhile. For drawings which merely communicate with shops, it is probably better to use conventional paper and pencil." Over time efforts would be directed toward the goal of having the designers drawings communicate not just with shops but with the shop tool itself. This goal would be a long time arriving.

The first commercial applications of CAD were in large companies in the automotive and aerospace industries, as well as in electronics. Only large corporations could afford the computers capable of performing the calculations. Notable company projects were, a joint project of GM (Dr. Patrick J.Hanratty) and IBM (Sam Matsa, Doug Ross`s MIT APT research assistant) to develop a prototype system for design engineers DAC-1 (Design Augmented by Computer) 1964; Lockheed projects; Bell GRAPHIC 1 and Renault.

One of the most influential events in the development of CAD was the founding of MCS (Manufacturing and Consulting Services Inc.) in 1971 by Dr. P. J. Hanratty, who wrote the system ADAM (Automated Drafting And Machining) but more importantly supplied code to companies such as McDonnell Douglas (Unigraphics), Computervision (CADDS), Calma, Gerber, Autotrol and Control Data.

As computers became more affordable, the application areas have gradually expanded. The development of CAD software for personal desktop computers was the impetus for almost universal application in all areas of construction.

Other key points in the 1960s and 1970s would be the foundation of CAD systems United Computing, Intergraph, IBM, Intergraph IGDS in 1974 (which led to Bentley Systems MicroStation in 1984).

CAD implementations have evolved dramatically since then. Initially, with 3D in the 1970s, it was typically limited to producing drawings similar to hand-drafted drawings. Advances in programming and computer hardware, notably solid modeling in the 1980s, have allowed more versatile applications of computers in design activities.

Key products for 1981 were the solid modelling packages - Romulus (ShapeData) and Uni-Solid (Unigraphics) based on PADL-2 and the release of the surface modeler CATIA (Dassault Systemes). Autodesk was founded 1982 by John Walker, which led to the 2D system AutoCAD. The next milestone was the release of Pro/ENGINEER in 1987, which heralded greater usage of feature-based modeling methods and parametric linking of the parameters of features. Also of importance to the development of CAD was the development of the B-rep solid modeling kernels (engines for manipulating geometrically and topologically consistent 3D objects) Parasolid (ShapeData) and ACIS (Spatial Technology Inc.) at the end of the 1980s and beginning of the 1990s, both inspired by the work of Ian Braid. This led to the release of mid-range packages such as SolidWorks and Tri-

Spective (later known as IRONCAD) in 1995, Solid Edge (then Intergraph) in 1996 and Autodesk Inventor in 1999. An independent geometric modeling kernel has been evolving in Russia since the 1990s. Nikolay Golovanov joined ASCON Company in 1994 from the Kolomna Engineering Design Bureau and began development of C3D – the geometric kernel of the Russian popular CAD system, KOMPAS-3D. Nowadays, C3D (C3D Labs) is the most valued Russian CAD product in the category of "components", i.e. products designed for integration in the end-user CAD systems of Russian and global vendors.

Computer-aided Manufacturing

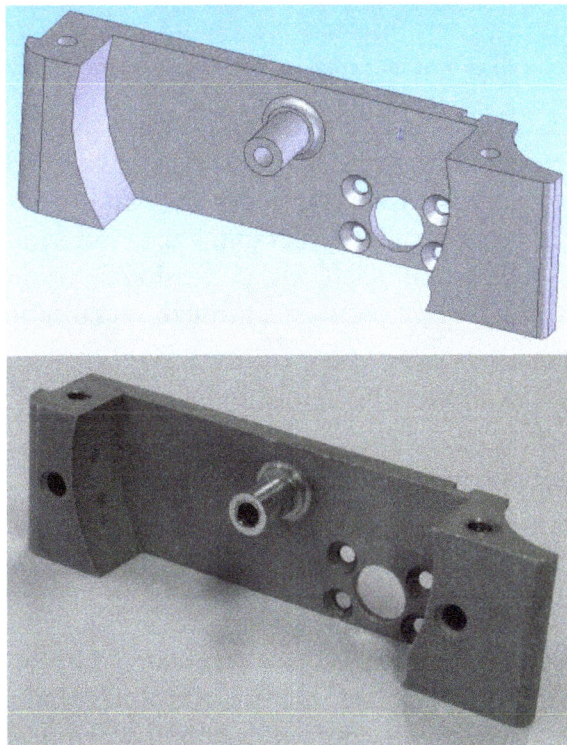

CAD model and CNC machined part

Computer-aided manufacturing (CAM) is the use of software to control machine tools and related ones in the manufacturing of workpieces. This is not the only definition for CAM, but it is the most common; CAM may also refer to the use of a computer to assist in all operations of a manufacturing plant, including planning, management, transportation and storage. Its primary purpose is to create a faster production process and components and tooling with more precise dimensions and material consistency, which in some cases, uses only the required amount of raw material (thus minimizing waste), while simultaneously reducing energy consumption. CAM is now a system used in schools and lower educational purposes. CAM is a subsequent computer-aided process after computer-aided design (CAD) and sometimes computer-aided engineering (CAE), as the model generated in CAD and verified in CAE can be input into CAM software, which then controls the machine tool. CAM is used in many schools alongside computer-aided design (CAD) to create objects.

Overview

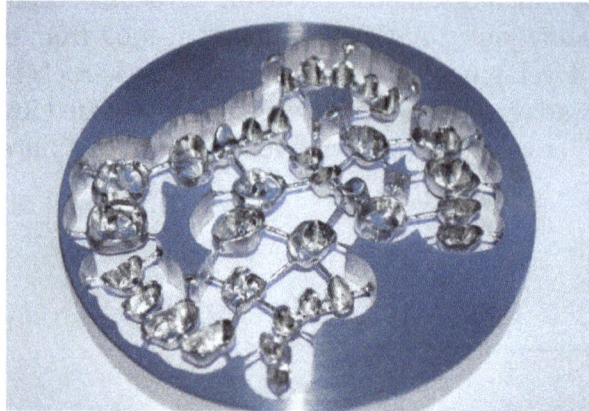

Chrome-cobalt disc with crowns for dental implants, manufactured using WorkNC CAM

Traditionally, CAM has been considered as a numerical control (NC) programming tool, where in two-dimensional (2-D) or three-dimensional (3-D) models of components generated in CADAs with other "Computer-Aided" technologies, CAM does not eliminate the need for skilled professionals such as manufacturing engineers, NC programmers, or machinists. CAM, in fact, leverages both the value of the most skilled manufacturing professionals through advanced productivity tools, while building the skills of new professionals through visualization, simulation and optimization tools.

History

Early commercial applications of CAM was in large companies in the automotive and aerospace industries, for example Pierre Béziers work developing the CAD/CAM application UNISURF in the 1960s for car body design and tooling at Renault.

Historically, CAM software was seen to have several shortcomings that necessitated an overly high level of involvement by skilled CNC machinists. Fallows created the first CAD software but this had severe shortcomings and was promptly taken back into the developing stage. CAM software would output code for the least capable machine, as each machine tool control added on to the standard G-code set for increased flexibility. In some cases, such as improperly set up CAM software or specific tools, the CNC machine required manual editing before the program will run properly. None of these issues were so insurmountable that a thoughtful engineer or skilled machine operator could not overcome for prototyping or small production runs; G-Code is a simple language. In high production or high precision shops, a different set of problems were encountered where an experienced CNC machinist must both hand-code programs and run CAM software.

Integration of CAD with other components of CAD/CAM/CAE Product lifecycle management (PLM) environment requires an effective CAD data exchange. Usually it had been necessary to force the CAD operator to export the data in one of the common data formats, such as IGES or STL or Parasolid formats that are supported by a wide variety of software. The output from the CAM software is usually a simple text file of G-code/M-codes, sometimes many thousands of commands long, that is then transferred to a machine tool using a direct numerical control (DNC) program or in modern Controllers using a common USB Storage Device.

CAM packages could not, and still cannot, reason as a machinist can. They could not optimize toolpaths to the extent required of mass production. Users would select the type of tool, machining process and paths to be used. While an engineer may have a working knowledge of G-code programming, small optimization and wear issues compound over time. Mass-produced items that require machining are often initially created through casting or some other non-machine method. This enables hand-written, short, and highly optimized G-code that could not be produced in a CAM package.

At least in the United States, there is a shortage of young, skilled machinists entering the workforce able to perform at the extremes of manufacturing; high precision and mass production. As CAM software and machines become more complicated, the skills required of a machinist or machine operator advance to approach that of a computer programmer and engineer rather than eliminating the CNC machinist from the workforce.

Typical Areas of Concern:

- High Speed Machining, including streamlining of tool paths
- Multi-function Machining
- 5 Axis Machining
- Feature recognition and machining
- Automation of Machining processes
- Ease of Use

Overcoming Historical Shortcomings

Over time, the historical shortcomings of CAM are being attenuated, both by providers of niche solutions and by providers of high-end solutions. This is occurring primarily in three arenas:

- Ease of usage
- Manufacturing complexity
- Integration with PLM and the extended enterprise

Ease in use

> For the user who is just getting started as a CAM user, out-of-the-box capabilities providing Process Wizards, templates, libraries, machine tool kits, automated feature based machining and job function specific tailorable user interfaces build user confidence and speed the learning curve.

> User confidence is further built on 3D visualization through a closer integration with the 3D CAD environment, including error-avoiding simulations and optimizations.

Manufacturing complexity

> The manufacturing environment is increasingly complex. The need for CAM and PLM tools

by buMs are NC programmer or machinist is similar to the need for computer assistance by the pilot of modern aircraft systems. The modern machinery cannot be properly used without this assistance.

Today's CAM systems support the full range of machine tools including: turning, 5 axis machining and wire EDM. Today's CAM user can easily generate streamlined tool paths, optimized tool axis tilt for higher feed rates, better tool life and surface finish and optimized Z axis depth cuts as well as driving non-cutting operations such as the specification of probing motions.

Integration with PLM and the extended enterpriseLM to integrate manufacturing with enterprise operations from concept through field support of the finished product.

To ensure ease of use appropriate to user objectives, modern CAM solutions are scalable from a stand-alone CAM system to a fully integrated multi-CAD 3D solution-set. These solutions are created to meet the full needs of manufacturing personnel including part planning, shop documentation, resource management and data management and exchange. To prevent these solutions from detailed tool specific information a dedicated tool management

Machining Process

Most machining progresses through many stages, each of which is implemented by a variety of basic and sophisticated strategies, depending on the material and the software available.

Roughing

This process begins with raw stock, known as billet, and cuts it very roughly to shape of the final model. In milling, the result often gives the appearance of terraces, because the strategy has taken advantage of the ability to cut the model horizontally. Common strategies are zig-zag clearing, offset clearing, plunge roughing, rest-roughing.

Semi-f

This process begins with a roughed part that unevenly approximates the model and cuts to within a fixed offset distance from the model. The semi-finishing pass must leave a small amount of material so the tool can cut accurately while finishing, but not so little that the tool and material deflect instead of sending. Common strategies are raster passes, waterline passes, constant step-over passes, pencil milling.

Finishing

Finishing involves a slow pass across the material in very fine steps to produce the finished part. In finishing, the step between one pass and another is minimal. Feed rates are low and spindle speeds are raised to produce an accurate surface.

Contour Milling

In milling applications on hardware with five or more axes, a separate finishing process called contouring can be performed. Instead of stepping down in fine-grained increments to approximate a surface, the work piece is rotated to make the cutting surfaces of the tool tangent to the ideal part features. This produces an excellent surface finish with high dimensional accuracy.

Computational Fluid Dynamics

Computational fluid dynamics (CFD) is a branch of fluid mechanics that uses numerical analysis and algorithms to solve and analyze problems that involve fluid flows. Computers are used to perform the calculations required to simulate the interaction of liquids and gases with surfaces defined by boundary conditions. With high-speed supercomputers, better solutions can be achieved. Ongoing research yields software that improves the accuracy and speed of complex simulation scenarios such as transonic or turbulent flows. Initial experimental validation of such software is performed using a wind tunnel with the final validation coming in full-scale testing, e.g. flight tests.

Background and History

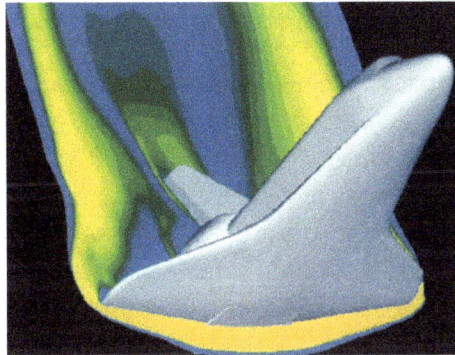

A computer simulation of high velocity air flow around the Space Shuttle during re-entry.

A simulation of the Hyper-X scramjet vehicle in operation at Mach-7

The fundamental basis of almost all CFD problems are the Navier–Stokes equations, which define many single-phase (gas or liquid, but not both) fluid flows. These equations can be simplified by removing terms describing viscous actions to yield the Euler equations. Further simplification, by removing terms describing vorticity yields the full potential equations. Finally, for small perturbations in subsonic and supersonic flows (not transonic or hypersonic) these equations can be linearized to yield the linearized potential equations.

Historically, methods were first developed to solve the linearized potential equations. Two-dimensional (2D) methods, using conformal transformations of the flow about a cylinder to the flow about an airfoil were developed in the 1930s.

One of the earliest type of calculations resembling modern CFD are those by Lewis Fry Richardson, in the sense that these calculations used finite differences and divided the physical space in cells. Although they failed dramatically, these calculations, together with Richardson's book "Weather prediction by numerical process", set the basis for modern CFD and numerical meteorology. In fact, early CFD calculations during the 1940s using ENIAC used methods close to those in Richardson's 1922 book.

The computer power available paced development of three-dimensional methods. Probably the first work using computers to model fluid flow, as governed by the Navier-Stokes equations, was performed at Los Alamos National Lab, in the T3 group. This group was led by Francis H. Harlow, who is widely considered as one of the pioneers of CFD. From 1957 to late 1960s, this group developed a variety of numerical methods to simulate transient two-dimensional fluid flows, such as Particle-in-cell method (Harlow, 1957), Fluid-in-cell method (Gentry, Martin and Daly, 1966), Vorticity stream function method (Jake Fromm, 1963), and Marker-and-cell method (Harlow and Welch, 1965). Fromm's vorticity-stream-function method for 2D, transient, incompressible flow was the first treatment of strongly contorting incompressible flows in the world.

The first paper with three-dimensional model was published by John Hess and A.M.O. Smith of Douglas Aircraft in 1967. This method discretized the surface of the geometry with panels, giving rise to this class of programs being called Panel Methods. Their method itself was simplified, in that it did not include lifting flows and hence was mainly applied to ship hulls and aircraft fuselages. The first lifting Panel Code (A230) was described in a paper written by Paul Rubbert and Gary Saaris of Boeing Aircraft in 1968. In time, more advanced three-dimensional Panel Codes were developed at Boeing (PANAIR, A502), Lockheed (Quadpan), Douglas (HESS), McDonnell Aircraft (MACAERO), NASA (PMARC) and Analytical Methods (WBAERO, USAERO and VSAERO). Some (PANAIR, HESS and MACAERO) were higher order codes, using higher order distributions of surface singularities, while others (Quadpan, PMARC, USAERO and VSAERO) used single singularities on each surface panel. The advantage of the lower order codes was that they ran much faster on the computers of the time. Today, VSAERO has grown to be a multi-order code and is the most widely used program of this class. It has been used in the development of many submarines, surface ships, automobiles, helicopters, aircraft, and more recently wind turbines. Its sister code, USAERO is an unsteady panel method that has also been used for modeling such things as high speed trains and racing yachts. The NASA PMARC code from an early version of VSAERO and a derivative of PMARC, named CMARC, is also commercially available.

In the two-dimensional realm, a number of Panel Codes have been developed for airfoil analysis and design. The codes typically have a boundary layer analysis included, so that viscous effects can be modeled. Professor Richard Eppler of the University of Stuttgart developed the PROFILE code, partly with NASA funding, which became available in the early 1980s. This was soon followed by MIT Professor Mark Drela's XFOIL code. Both PROFILE and XFOIL incorporate two-dimensional panel codes, with coupled boundary layer codes for airfoil analysis work. PROFILE uses a conformal transformation method for inverse airfoil design, while XFOIL has both a conformal transformation and an inverse panel method for airfoil design.

An intermediate step between Panel Codes and Full Potential codes were codes that used the Transonic Small Disturbance equations. In particular, the three-dimensional WIBCO code, developed by Charlie Boppe of Grumman Aircraft in the early 1980s has seen heavy use.

Developers turned to Full Potential codes, as panel methods could not calculate the non-linear flow present at transonic speeds. The first description of a means of using the Full Potential equations was published by Earll Murman and Julian Cole of Boeing in 1970. Frances Bauer, Paul Garabedian and David Korn of the Courant Institute at New York University (NYU) wrote a series of two-dimensional Full Potential airfoil codes that were widely used, the most important being named Program H. A further growth of Program H was developed by Bob Melnik and his group at Grumman Aerospace as Grumfoil. Antony Jameson, originally at Grumman Aircraft and the Courant Institute of NYU, worked with David Caughey to develop the important three-dimensional Full Potential code FLO22 in 1975. Many Full Potential codes emerged after this, culminating in Boeing's Tranair (A633) code, which still sees heavy use.

The next step was the Euler equations, which promised to provide more accurate solutions of transonic flows. The methodology used by Jameson in his three-dimensional FLO57 code (1981) was used by others to produce such programs as Lockheed's TEAM program and IAI/Analytical Methods' MGAERO program. MGAERO is unique in being a structured cartesian mesh code, while most other such codes use structured body-fitted grids (with the exception of NASA's highly successful CART3D code, Lockheed's SPLITFLOW code and Georgia Tech's NASCART-GT). Antony Jameson also developed the three-dimensional AIRPLANE code which made use of unstructured tetrahedral grids.

In the two-dimensional realm, Mark Drela and Michael Giles, then graduate students at MIT, developed the ISES Euler program (actually a suite of programs) for airfoil design and analysis. This code first became available in 1986 and has been further developed to design, analyze and optimize single or multi-element airfoils, as the MSES program. MSES sees wide use throughout the world. A derivative of MSES, for the design and analysis of airfoils in a cascade, is MISES, developed by Harold "Guppy" Youngren while he was a graduate student at MIT.

The Navier–Stokes equations were the ultimate target of development. Two-dimensional codes, such as NASA Ames' ARC2D code first emerged. A number of three-dimensional codes were developed (ARC3D, OVERFLOW, CFL3D are three successful NASA contributions), leading to numerous commercial packages.

Methodology

In all of these approaches the same basic procedure is followed.

- During preprocessing
    - The geometry (physical bounds) of the problem is defined.
    - The volume occupied by the fluid is divided into discrete cells (the mesh). The mesh may be uniform or non-uniform.
    - The physical modeling is defined – for example, the equations of motion + enthalpy + radiation + species conservation

- Boundary conditions are defined. This involves specifying the fluid behaviour and properties at the boundaries of the problem. For transient problems, the initial conditions are also defined.

- The simulation is started and the equations are solved iteratively as a steady-state or transient.

- Finally a postprocessor is used for the analysis and visualization of the resulting solution.

Discretization Methods

The stability of the selected discretisation is generally established numerically rather than analytically as with simple linear problems. Special care must also be taken to ensure that the discretisation handles discontinuous solutions gracefully. The Euler equations and Navier–Stokes equations both admit shocks, and contact surfaces.

Some of the discretization methods being used are:

Finite Volume Method

The finite volume method (FVM) is a common approach used in CFD codes, as it has an advantage in memory usage and solution speed, especially for large problems, high Reynolds number turbulent flows, and source term dominated flows (like combustion).

In the finite volume method, the governing partial differential equations (typically the Navier-Stokes equations, the mass and energy conservation equations, and the turbulence equations) are recast in a conservative form, and then solved over discrete control volumes. This discretization guarantees the conservation of fluxes through a particular control volume. The finite volume equation yields governing equations in the form,

$$\frac{\partial}{\partial t} \iiint Q dV + \iint F d\mathbf{A} = 0,$$

where Q is the vector of conserved variables, F is the vector of fluxes, V is the volume of the control volume element, and $\mathbf{A}$ is the surface area of the control volume element.

Finite Element Method

The finite element method (FEM) is used in structural analysis of solids, but is also applicable to fluids. However, the FEM formulation requires special care to ensure a conservative solution. The FEM formulation has been adapted for use with fluid dynamics governing equations. Although FEM must be carefully formulated to be conservative, it is much more stable than the finite volume approach. However, FEM can require more memory and has slower solution times than the FVM.

In this method, a weighted residual equation is formed:

$$R_i = \iiint W_i Q dV^e$$

where R_i is the equation residual at an element vertex i, Q is the conservation equation expressed on an element basis, W_i is the weight factor, and V^e is the volume of the element.

Finite Difference Method

The finite difference method (FDM) has historical importance and is simple to program. It is currently only used in few specialized codes, which handle complex geometry with high accuracy and efficiency by using embedded boundaries or overlapping grids (with the solution interpolated across each grid).

$$\frac{\partial Q}{\partial t} + \frac{\partial F}{\partial x} + \frac{\partial G}{\partial y} + \frac{\partial H}{\partial z} = 0$$

where Q is the vector of conserved variables, and F G, and H, are the fluxes in the x, y, and z directions respectively.

Spectral Element Method

Spectral element method is a finite element type method. It requires the mathematical problem (the partial differential equation) to be cast in a weak formulation. This is typically done by multiplying the differential equation by an arbitrary test function and integrating over the whole domain. Purely mathematically, the test functions are completely arbitrary - they belong to an infinite-dimensional function space. Clearly an infinite-dimensional function space cannot be represented on a discrete spectral element mesh; this is where the spectral element discretization begins. The most crucial thing is the choice of interpolating and testing functions. In a standard, low order FEM in 2D, for quadrilateral elements the most typical choice is the bilinear test or interpolating function of the form $v(x, y) = ax + by + cxy + d$. In a spectral element method however, the interpolating and test functions are chosen to be polynomials of a very high order (typically e.g. of the 10th order in CFD applications). This guarantees the rapid convergence of the method. Furthermore, very efficient integration procedures must be used, since the number of integrations to be performed in a numerical codes is big. Thus, high order Gauss integration quadratures are employed, since they achieve the highest accuracy with the smallest number of computations to be carried out. At the time there are some academic CFD codes based on the spectral element method and some more are currently under development, since the new time-stepping schemes arise in the scientific world. You can refer to the C-CFD website to see movies of incompressible flows in channels simulated with a spectral element solver or to the Numerical Mechanics website to see a movie of the lid-driven cavity flow obtained with a compeletely novel unconditionally stable time-stepping scheme combined with a spectral element solver.

Boundary Element Method

In the boundary element method, the boundary occupied by the fluid is divided into a surface mesh.

High-resolution Discretization Schemes

High-resolution schemes are used where shocks or discontinuities are present. Capturing sharp changes in the solution requires the use of second or higher-order numerical schemes that do not introduce spurious oscillations. This usually necessitates the application of flux limiters to ensure that the solution is total variation diminishing.

Turbulence Models

In computational modeling of turbulent flows, one common objective is to obtain a model that can predict quantities of interest, such as fluid velocity, for use in engineering designs of the system being modeled. For turbulent flows, the range of length scales and complexity of phenomena involved in turbulence make most modeling approaches prohibitively expensive; the resolution required to resolve all scales involved in turbulence is beyond what is computationally possible. The primary approach in such cases is to create numerical models to approximate unresolved phenomena. This section lists some commonly used computational models for turbulent flows.

Turbulence models can be classified based on computational expense, which corresponds to the range of scales that are modeled versus resolved (the more turbulent scales that are resolved, the finer the resolution of the simulation, and therefore the higher the computational cost). If a majority or all of the turbulent scales are not modeled, the computational cost is very low, but the tradeoff comes in the form of decreased accuracy.

In addition to the wide range of length and time scales and the associated computational cost, the governing equations of fluid dynamics contain a non-linear convection term and a non-linear and non-local pressure gradient term. These nonlinear equations must be solved numerically with the appropriate boundary and initial conditions.

Reynolds-Averaged Navier–Stokes

Reynolds-averaged Navier-Stokes (RANS) equations are the oldest approach to turbulence modeling. An ensemble version of the governing equations is solved, which introduces new *apparent stresses* known as Reynolds stresses. This adds a second order tensor of unknowns for which various models can provide different levels of closure. It is a common misconception that the RANS equations do not apply to flows with a time-varying mean flow because these equations are 'time-averaged'. In fact, statistically unsteady (or non-stationary) flows can equally be treated. This is sometimes referred to as URANS. There is nothing inherent in Reynolds averaging to preclude this, but the turbulence models used to close the equations are valid only as long as the time over which these changes in the mean occur is large compared to the time scales of the turbulent motion containing most of the energy.

RANS models can be divided into two broad approaches:

Boussinesq hypothesis

> This method involves using an algebraic equation for the Reynolds stresses which include determining the turbulent viscosity, and depending on the level of sophistication of the model, solving transport equations for determining the turbulent kinetic energy and dissi-

pation. Models include k-ε (Launder and Spalding), Mixing Length Model (Prandtl), and Zero Equation Model (Cebeci and Smith). The models available in this approach are often referred to by the number of transport equations associated with the method. For example, the Mixing Length model is a "Zero Equation" model because no transport equations are solved; the is a "Two Equation" model because two transport equations (one for and one for) are solved.

Reynolds stress model (RSM)

This approach attempts to actually solve transport equations for the Reynolds stresses. This means introduction of several transport equations for all the Reynolds stresses and hence this approach is much more costly in CPU effort.

Large Eddy Simulation

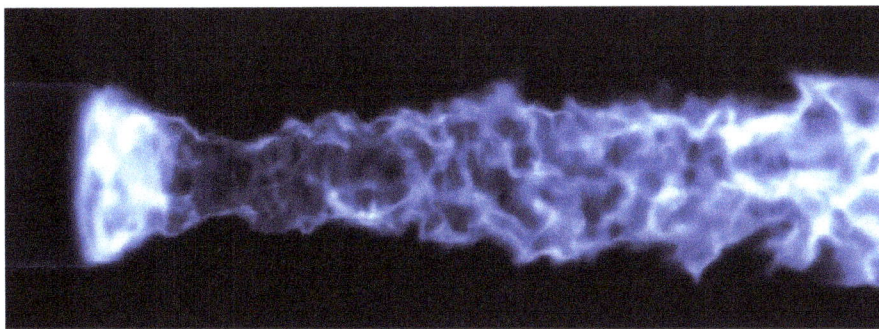

Volume rendering of a non-premixed swirl flame as simulated by LES.

Large eddy simulation (LES) is a technique in which the smallest scales of the flow are removed through a filtering operation, and their effect modeled using subgrid scale models. This allows the largest and most important scales of the turbulence to be resolved, while greatly reducing the computational cost incurred by the smallest scales. This method requires greater computational resources than RANS methods, but is far cheaper than DNS.

Detached Eddy Simulation

Detached eddy simulations (DES) is a modification of a RANS model in which the model switches to a subgrid scale formulation in regions fine enough for LES calculations. Regions near solid boundaries and where the turbulent length scale is less than the maximum grid dimension are assigned the RANS mode of solution. As the turbulent length scale exceeds the grid dimension, the regions are solved using the LES mode. Therefore, the grid resolution for DES is not as demanding as pure LES, thereby considerably cutting down the cost of the computation. Though DES was initially formulated for the Spalart-Allmaras model (Spalart et al., 1997), it can be implemented with other RANS models (Strelets, 2001), by appropriately modifying the length scale which is explicitly or implicitly involved in the RANS model. So while Spalart-Allmaras model based DES acts as LES with a wall model, DES based on other models (like two equation models) behave as a hybrid RANS-LES model. Grid generation is more complicated than for a simple RANS or LES case due to the RANS-LES switch. DES is a non-zonal approach and provides a single smooth velocity field across the RANS and the LES regions of the solutions.

Direct Numerical Simulation

Direct numerical simulation (DNS) resolves the entire range of turbulent length scales. This marginalizes the effect of models, but is extremely expensive. The computational cost is proportional to Re^3. DNS is intractable for flows with complex geometries or flow configurations.

Coherent vortex simulation

The coherent vortex simulation approach decomposes the turbulent flow field into a coherent part, consisting of organized vortical motion, and the incoherent part, which is the random background flow. This decomposition is done using wavelet filtering. The approach has much in common with LES, since it uses decomposition and resolves only the filtered portion, but different in that it does not use a linear, low-pass filter. Instead, the filtering operation is based on wavelets, and the filter can be adapted as the flow field evolves. Farge and Schneider tested the CVS method with two flow configurations and showed that the coherent portion of the flow exhibited the $-\dfrac{40}{39}$ energy spectrum exhibited by the total flow, and corresponded to coherent structures (vortex tubes), while the incoherent parts of the flow composed homogeneous background noise, which exhibited no organized structures. Goldstein and Vasilyev applied the FDV model to large eddy simulation, but did not assume that the wavelet filter completely eliminated all coherent motions from the subfilter scales. By employing both LES and CVS filtering, they showed that the SFS dissipation was dominated by the SFS flow field's coherent portion.

PDF Methods

Probability density function (PDF) methods for turbulence, first introduced by Lundgren, are based on tracking the one-point PDF of the velocity, $f_V(v; x, t)dv$, which gives the probability of the velocity at point x being between v and $v + dv$. This approach is analogous to the kinetic theory of gases, in which the macroscopic properties of a gas are described by a large number of particles. PDF methods are unique in that they can be applied in the framework of a number of different turbulence models; the main differences occur in the form of the PDF transport equation. For example, in the context of large eddy simulation, the PDF becomes the filtered PDF. PDF methods can also be used to describe chemical reactions, and are particularly useful for simulating chemically reacting flows because the chemical source term is closed and does not require a model. The PDF is commonly tracked by using Lagrangian particle methods; when combined with large eddy simulation, this leads to a Langevin equation for subfilter particle evolution.

Vortex Method

The vortex method is a grid-free technique for the simulation of turbulent flows. It uses vortices as the computational elements, mimicking the physical structures in turbulence. Vortex methods were developed as a grid-free methodology that would not be limited by the fundamental smoothing effects associated with grid-based methods. To be practical, however, vortex methods require means for rapidly computing velocities from the vortex elements – in other words they require the solution to a particular form of the N-body problem (in which the motion of N objects is tied to their mutual influences). A breakthrough came in the late 1980s with the development of the fast multipole method (FMM), an algorithm by V. Rokhlin (Yale) and L. Greengard (Courant Institute). This breakthrough paved the way to practical computation of the velocities from the vortex elements and is the basis of successful algorithms. They are especially well-suited to simulating filamentary motion, such as wisps of smoke, in real-time simulations such as video games, because of the fine detail achieved using minimal computation.

Software based on the vortex method offer a new means for solving tough fluid dynamics problems with minimal user intervention.All that is required is specification of problem geometry and setting of boundary and initial conditions. Among the significant advantages of this modern technology;

- It is practically grid-free, thus eliminating numerous iterations associated with RANS and LES.

- All problems are treated identically. No modeling or calibration inputs are required.

- Time-series simulations, which are crucial for correct analysis of acoustics, are possible.

- The small scale and large scale are accurately simulated at the same time.

Vorticity Confinement Method

The vorticity confinement (VC) method is an Eulerian technique used in the simulation of turbulent wakes. It uses a solitary-wave like approach to produce a stable solution with no numerical spreading. VC can capture the small-scale features to within as few as 2 grid cells. Within these features, a nonlinear difference equation is solved as opposed to the finite difference equation. VC is similar to shock capturing methods, where conservation laws are satisfied, so that the essential integral quantities are accurately computed.

Linear Eddy Model

The Linear eddy model is a technique used to simulate the convective mixing that takes place in turbulent flow. Specifically, it provides a mathematical way to describe the interactions of a scalar variable within the vector flow field. It is primarily used in one-dimensional representations of turbulent flow, since it can be applied across a wide range of length scales and Reynolds numbers. This model is generally used as a building block for more complicated flow representations, as it provides high resolution predictions that hold across a large range of flow conditions.

Two-phase Flow

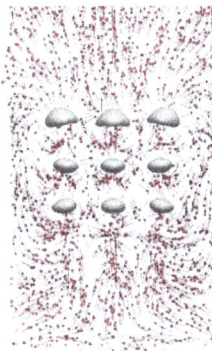

Simulation of bubble swarm using volume of fluid method

The modeling of two-phase flow is still under development. Different methods have been proposed lately. The Volume of fluid method has received a lot of attention lately, for problems that do not have dispersed particles, but the Level set method and front tracking are also valuable approaches. Most of these methods are either good in maintaining a sharp interface or at conserving mass. This is crucial since the evaluation of the density, viscosity and surface tension is based on the values averaged over the interface. Lagrangian multiphase models, which are used for dispersed media, are based on solving the Lagrangian equation of motion for the dispersed phase.

Solution Algorithms

Discretization in the space produces a system of ordinary differential equations for unsteady problems and algebraic equations for steady problems. Implicit or semi-implicit methods are generally used to integrate the ordinary differential equations, producing a system of (usually) nonlinear algebraic equations. Applying a Newton or Picard iteration produces a system of linear equations which is nonsymmetric in the presence of advection and indefinite in the presence of incompressibility. Such systems, particularly in 3D, are frequently too large for direct solvers, so iterative methods are used, either stationary methods such as successive overrelaxation or Krylov subspace methods. Krylov methods such as GMRES, typically used with preconditioning, operate by minimizing the residual over successive subspaces generated by the preconditioned operator.

Multigrid has the advantage of asymptotically optimal performance on many problems. Traditional solvers and preconditioners are effective at reducing high-frequency components of the residual, but low-frequency components typically require many iterations to reduce. By operating on multiple scales, multigrid reduces all components of the residual by similar factors, leading to a mesh-independent number of iterations.

For indefinite systems, preconditioners such as incomplete LU factorization, additive Schwarz, and multigrid perform poorly or fail entirely, so the problem structure must be used for effective preconditioning. Methods commonly used in CFD are the SIMPLE and Uzawa algorithms which exhibit mesh-dependent convergence rates, but recent advances based on block LU factorization combined with multigrid for the resulting definite systems have led to preconditioners that deliver mesh-independent convergence rates.

Unsteady Aerodynamics

CFD made a major break through in late 70s with the introduction of LTRAN2, a 2-D code to model oscillating airfoils based on transonic small perturbation theory by Ballhaus and associates. It uses a Murman-Cole switch algorithm for modeling the moving shock-waves. Later it was extended to 3-D with use of a rotated difference scheme by AFWAL/Boeing that resulted in LTRAN3.

Biomedical Engineering

CFD investigations are used to clarify the characteristics of aortic flow in detail that are otherwise invisible to experimental measurements. To analyze these conditions, CAD models of the human vascular system are extracted employing modern imaging techniques. A 3D model is reconstructed from this data and the fluid flow can be computed. Blood properties like Non-Newtonian behavior and realistic boundary conditions (e.g. systemic pressure) have to be taken into consideration.

Therefore, making it possible to analyze and optimize the flow in the cardiovascular system for different applications.

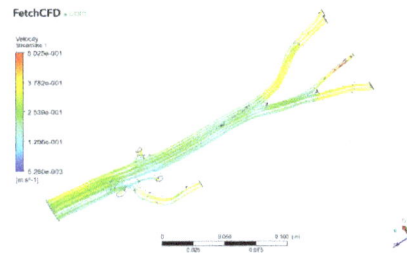

Simulation of blood flow in a human aorta

References

- Golovanov, Nikolay (2014). Geometric Modeling: The mathematics of shapes. CreateSpace Independent Publishing Platform (December 24, 2014). p. Back cover. ISBN 978-1497473195.

- Narayan, K. Lalit (2008). Computer Aided Design and Manufacturing. New Delhi: Prentice Hall of India. p. 4. ISBN 812033342X.

- Duggal, Vijay (2000). Cadd Primer: A General Guide to Computer Aided Design and Drafting-Cadd, CAD. Mailmax Pub. ISBN 978-0962916595.

- Farin, Gerald; Hoschek, Josef; Kim, Myung-Soo (2002). Handbook of computer aided geometric design [electronic resource]. Elsevier. ISBN 978-0-444-51104-1.

- Pottmann, H.; Brell-Cokcan, S. and Wallner, J. (2007) "Discrete surfaces for architectural design", pp. 213–234 in Curve and Surface Design, Patrick Chenin, Tom Lyche and Larry L. Schumaker (eds.), Nashboro Press, ISBN 978-0-9728482-7-5.

- Hurst, J. (1989) Retrospectives II: The Early Years in Computer Graphics, pp. 39–73 in SIGGRAPH 89 Panel Proceedings, ACM New York, NY, USA, ISBN 0-89791-353-1 doi:10.1145/77276.77280

- Golovanov, Nikolay (December 24, 2014). Geometric Modeling: The mathematics of shapes. CreateSpace Independent Publishing Platform (December 24, 2014). p. Back cover. ASIN 1497473195. ISBN 978-1497473195.

- U.S. Congress, Office of Technology Assessment (1984). Computerized manufacturing automation. DIANE Publishing. p. 48. ISBN 978-1-4289-2364-5.

- Hosking, Dian Marie; Anderson, Neil (1992), Organizational change and innovation, Taylor & Francis, p. 240, ISBN 978-0-415-06314-2

- Matthews, Clifford (2005). Aeronautical engineer's data book (2nd ed.). Butterworth-Heinemann. p. 229. ISBN 978-0-7506-5125-7.

- Pichler, Franz; Moreno-Díaz, Roberto (1992). Computer aided systems theory. Springer. p. 602. ISBN 978-3-540-55354-0.

- Boothroyd, Geoffrey; Knight, Winston Anthony (2006). Fundamentals of machining and machine tools (3rd ed.). CRC Press. p. 401. ISBN 978-1-57444-659-3.

- Bauer, F., Garabedian, P., and Korn, D. G., "A Theory of Supercritical Wing Sections, with Computer Programs and Examples," Lecture Notes in Economics and Mathematical Systems 66, Springer-Verlag, May 1972. ISBN 978-3540058076

- Patankar, Suhas V. (1980). Numerical Heat Transfer and Fluid FLow. Hemisphere Publishing Corporation. ISBN 0891165223.

- Fox, Rodney (2003). Computational models for turbulent reacting flows. Cambridge University Press. ISBN 978-0-521-65049-6.

4

Various Models of Mechanical Engineering

Mechanical engineering uses various design models upon which new creations are based. Some of the basic models and ideas used in this field are kinematic chain, rigid body dynamics, equations of motion etc. This chapter explains the fundamentals of mechanical engineering in an easy and comprehensive manner.

Multidisciplinary Design Optimization

Multi-disciplinary design optimization (MDO) is a field of engineering that uses optimization methods to solve design problems incorporating a number of disciplines. It is also known as multidisciplinary optimization and multidisciplinary system design optimization (MSDO).

MDO allows designers to incorporate all relevant disciplines simultaneously. The optimum of the simultaneous problem is superior to the design found by optimizing each discipline sequentially, since it can exploit the interactions between the disciplines. However, including all disciplines simultaneously significantly increases the complexity of the problem.

These techniques have been used in a number of fields, including automobile design, naval architecture, electronics, architecture, computers, and electricity distribution. However, the largest number of applications have been in the field of aerospace engineering, such as aircraft and spacecraft design. For example, the proposed Boeing blended wing body (BWB) aircraft concept has used MDO extensively in the conceptual and preliminary design stages. The disciplines considered in the BWB design are aerodynamics, structural analysis, propulsion, control theory, and economics.

History

Traditionally engineering has normally been performed by teams, each with expertise in a specific discipline, such as aerodynamics or structures. Each team would use its members' experience and judgement to develop a workable design, usually sequentially. For example, the aerodynamics experts would outline the shape of the body, and the structural experts would be expected to fit their design within the shape specified. The goals of the teams were generally performance-related, such as maximum speed, minimum drag, or minimum structural weight.

Between 1970 and 1990, two major developments in the aircraft industry changed the approach of aircraft design engineers to their design problems. The first was computer-aided design, which allowed designers to quickly modify and analyse their designs. The second was changes in the procurement policy of most airlines and military organizations, particularly the military of the United States, from a performance-centred approach to one that emphasized lifecycle cost issues. This led to an increased concentration on economic factors and the attributes known as the "ilities" including manufacturability, reliability, maintainability, etc.

Since 1990, the techniques have expanded to other industries. Globalization has resulted in more distributed, decentralized design teams. The high-performance personal computer has largely replaced the centralized supercomputer and the Internet and local area networks have facilitated sharing of design information. Disciplinary design software in many disciplines (such as OptiStruct or NASTRAN, a finite element analysis program for structural design) have become very mature. In addition, many optimization algorithms, in particular the population-based algorithms, have advanced significantly.

Origins in Structural Optimization

Whereas optimization methods are nearly as old as calculus, dating back to Isaac Newton, Leonhard Euler, Daniel Bernoulli, and Joseph Louis Lagrange, who used them to solve problems such as the shape of the catenary curve, numerical optimization reached prominence in the digital age. Its systematic application to structural design dates to its advocacy by Schmit in 1960. The success of structural optimization in the 1970s motivated the emergence of multidisciplinary design optimization (MDO) in the 1980s. Jaroslaw Sobieski championed decomposition methods specifically designed for MDO applications. The following synopsis focuses on optimization methods for MDO. First, the popular gradient-based methods used by the early structural optimization and MDO community are reviewed. Then those methods developed in the last dozen years are summarized.

Gradient-Based Methods

There were two schools of structural optimization practitioners using gradient-based methods during the 1960s and 1970s: optimality criteria and mathematical programming. The optimality criteria school derived recursive formulas based on the Karush–Kuhn–Tucker (KKT) necessary conditions for an optimal design. The KKT conditions were applied to classes of structural problems such as minimum weight design with constraints on stresses, displacements, buckling, or frequencies [Rozvany, Berke, Venkayya, Khot, et al.] to derive resizing expressions particular to each class. The mathematical programming school employed classical gradient-based methods to structural optimization problems. The method of usable feasible directions, Rosen's gradient projection (generalized reduce gradient) method, sequential unconstrained minimization techniques, sequential linear programming and eventually sequential quadratic programming methods were common choices. Schittkowski et al. reviewed the methods current by the early 1990s.

The gradient methods unique to the MDO community derive from the combination of optimality criteria with math programming, first recognized in the seminal work of Fleury and Schmit who constructed a framework of approximation concepts for structural optimization. They recognized that optimality criteria were so successful for stress and displacement constraints, because that approach amounted to solving the dual problem for Lagrange multipliers using linear Taylor series approximations in the reciprocal design space. In combination with other techniques to improve efficiency, such as constraint deletion, regionalization, and design variable linking, they succeeded in uniting the work of both schools. This approximation concepts based approach forms the basis of the optimization modules in modern structural design software such as Altair – Optistruct, ASTROS, MSC.Nastran, PHX ModelCenter, Genesis, iSight, and I-DEAS.

Approximations for structural optimization were initiated by the reciprocal approximation Schmit and Miura for stress and displacement response functions. Other intermediate variables were employed for plates. Combining linear and reciprocal variables, Starnes and Haftka developed a conservative approximation to improve buckling approximations. Fadel chose an appropriate intermediate design variable for each function based on a gradient matching condition for the previous point. Vanderplaats initiated a second generation of high quality approximations when he developed the force approximation as an intermediate response approximation to improve the approximation of stress constraints. Canfield developed a Rayleigh quotient approximation to improve the accuracy of eigenvalue approximations. Barthelemy and Haftka published a comprehensive review of approximations in 1993.

Non-Gradient-Based Methods

In recent years, non-gradient-based evolutionary methods including genetic algorithms, simulated annealing, and ant colony algorithms came into existence. At present, many researchers are striving to arrive at a consensus regarding the best modes and methods for complex problems like impact damage, dynamic failure, and real-time analyses. For this purpose, researchers often employ multiobjective and multicriteria design methods.

Recent MDO Methods

MDO practitioners have investigated optimization methods in several broad areas in the last dozen years. These include decomposition methods, approximation methods, evolutionary algorithms, memetic algorithms, response surface methodology, reliability-based optimization, and multi-objective optimization approaches.

The exploration of decomposition methods has continued in the last dozen years with the development and comparison of a number of approaches, classified variously as hierarchic and non hierarchic, or collaborative and non collaborative. Approximation methods spanned a diverse set of approaches, including the development of approximations based on surrogate models (often referred to as metamodels), variable fidelity models, and trust region management strategies. The development of multipoint approximations blurred the distinction with response surface methods. Some of the most popular methods include Kriging and the moving least squares method.

Response surface methodology, developed extensively by the statistical community, received much attention in the MDO community in the last dozen years. A driving force for their use has been the development of massively parallel systems for high performance computing, which are naturally suited to distributing the function evaluations from multiple disciplines that are required for the construction of response surfaces. Distributed processing is particularly suited to the design process of complex systems in which analysis of different disciplines may be accomplished naturally on different computing platforms and even by different teams.

Evolutionary methods led the way in the exploration of non-gradient methods for MDO applications. They also have benefited from the availability of massively parallel high performance computers, since they inherently require many more function evaluations than gradient-based methods. Their primary benefit lies in their ability to handle discrete design variables and the potential to find globally optimal solutions.

Reliability-based optimization (RBO) is a growing area of interest in MDO. Like response surface methods and evolutionary algorithms, RBO benefits from parallel computation, because the numeric integration to calculate the probability of failure requires many function evaluations. One of the first approaches employed approximation concepts to integrate the probability of failure. The classical first-order reliability method (FORM) and second-order reliability method (SORM) are still popular. Professor Ramana Grandhi used appropriate normalized variables about the most probable point of failure, found by a two-point adaptive nonlinear approximation to improve the accuracy and efficiency. Southwest Research Institute has figured prominently in the development of RBO, implementing state-of-the-art reliability methods in commercial software. RBO has reached sufficient maturity to appear in commercial structural analysis programs like Altair's Optistruct and MSC's Nastran.

Utility-based probability maximization (Bordley and Pollock, Operations Research, Sept, 2009, pg.1262) was developed in response to some logical concerns (e.g., Blau's Dilemma) with reliability-based design optimization. This approach focuses on maximizing the joint probability of both the objective function exceeding some value and of all the constraints being satisfied. When there is no objective function, utility-based probability maximization reduces to a probability-maximization problem. When there are no uncertainties in the constraints, it reduces to a constrained utility-maximization problem. (This second equivalence arises because the utility of a function can always be written as the probability of that function exceeding some random variable.) Because it changes the constrained optimization problem associated with reliability-based optimization into an unconstrained optimization problem, it often leads to computationally more tractable problem formulations.

In the marketing field there is a huge literature about optimal design for multiattribute products and services, based on experimental analisis to estimate models of consumers' utility functions. These methods are known as Conjoint Analysis. Respondents are presented with alternative products, measuring preferences about the alternatives using a variety of scales and the utility function is estimated with different methods (varying from regression and surface response methods to choice models). The best design is formulated after estimating the model. The experimental design is usually optimized to minimize the variance of the estimators. These methods are widely used in practice.

Problem formulation

Problem formulation is normally the most difficult part of the process. It is the selection of design variables, constraints, objectives, and models of the disciplines. A further consideration is the strength and breadth of the interdisciplinary coupling in the problem.

Design Variables

A design variable is a specification that is controllable from the point of view of the designer. For instance, the thickness of a structural member can be considered a design variable. Another might be the choice of material. Design variables can be continuous (such as a wing span), discrete (such as the number of ribs in a wing), or boolean (such as whether to build a monoplane or a biplane). Design problems with continuous variables are normally solved more easily.

Design variables are often bounded, that is, they often have maximum and minimum values. Depending on the solution method, these bounds can be treated as constraints or separately.

Constraints

A constraint is a condition that must be satisfied in order for the design to be feasible. An example of a constraint in aircraft design is that the lift generated by a wing must be equal to the weight of the aircraft. In addition to physical laws, constraints can reflect resource limitations, user requirements, or bounds on the validity of the analysis models. Constraints can be used explicitly by the solution algorithm or can be incorporated into the objective using Lagrange multipliers.

Objectives

An objective is a numerical value that is to be maximized or minimized. For example, a designer may wish to maximize profit or minimize weight. Many solution methods work only with single objectives. When using these methods, the designer normally weights the various objectives and sums them to form a single objective. Other methods allow multiobjective optimization, such as the calculation of a Pareto front.

Models

The designer must also choose models to relate the constraints and the objectives to the design variables. These models are dependent on the discipline involved. They may be empirical models, such as a regression analysis of aircraft prices, theoretical models, such as from computational fluid dynamics, or reduced-order models of either of these. In choosing the models the designer must trade off fidelity with analysis time.

The multidisciplinary nature of most design problems complicates model choice and implementation. Often several iterations are necessary between the disciplines in order to find the values of the objectives and constraints. As an example, the aerodynamic loads on a wing affect the structural deformation of the wing. The structural deformation in turn changes the shape of the wing and the aerodynamic loads. Therefore, in analysing a wing, the aerodynamic and structural analyses must be run a number of times in turn until the loads and deformation converge.

Standard Form

Once the design variables, constraints, objectives, and the relationships between them have been chosen, the problem can be expressed in the following form:

$$\text{find } \mathbf{x} \text{ that minimizes } J(\mathbf{x}) \text{ subject to } \mathbf{g}(\mathbf{x}) \leq \mathbf{0}, \; \mathbf{h}(\mathbf{x}) = \mathbf{0} \text{ and } \mathbf{x}_{lb} \leq \mathbf{x} \leq \mathbf{x}_{ub}$$

where J is an objective, $\mathbf{x}$ is a vector of design variables, $\mathbf{g}$ is a vector of inequality constraints, $\mathbf{h}$ is a vector of equality constraints, and $\mathbf{x}_{lb}$ and $\mathbf{x}_{ub}$ are vectors of lower and upper bounds on the design variables. Maximization problems can be converted to minimization problems by multiplying the objective by -1. Constraints can be reversed in a similar manner. Equality constraints can be replaced by two inequality constraints.

Problem Solution

The problem is normally solved using appropriate techniques from the field of optimization. These include gradient-based algorithms, population-based algorithms, or others. Very simple problems can sometimes be expressed linearly; in that case the techniques of linear programming are applicable.

Gradient-Based Methods

- Adjoint equation
- Newton's method
- Steepest descent
- Conjugate gradient
- Sequential quadratic programming

Gradient-Free Methods

- Hooke-Jeeves pattern search
- Nelder-Mead method

Population-Based Methods

- Genetic algorithm
- Memetic algorithm
- Particle swarm optimization
- Harmony search
- ODMA

Other Methods

- Random search
- Grid search
- Simulated annealing
- Direct search
- IOSO (Indirect Optimization based on Self-Organization)

Most of these techniques require large numbers of evaluations of the objectives and the constraints. The disciplinary models are often very complex and can take significant amounts of time for a single evaluation. The solution can therefore be extremely time-consuming. Many of the optimization techniques are adaptable to parallel computing. Much current research is focused on methods of decreasing the required time.

Also, no existing solution method is guaranteed to find the global optimum of a general problem. Gradient-based methods find local optima with high reliability but are normally unable to escape a local optimum. Stochastic methods, like simulated annealing and genetic algorithms, will find a good solution with high probability, but very little can be said about the mathematical properties of the solution. It is not guaranteed to even be a local optimum. These methods often find a different design each time they are run.

Kinematic Chain

The JPL mobile robot ATHLETE is a platform with six serial chain legs ending in wheels.

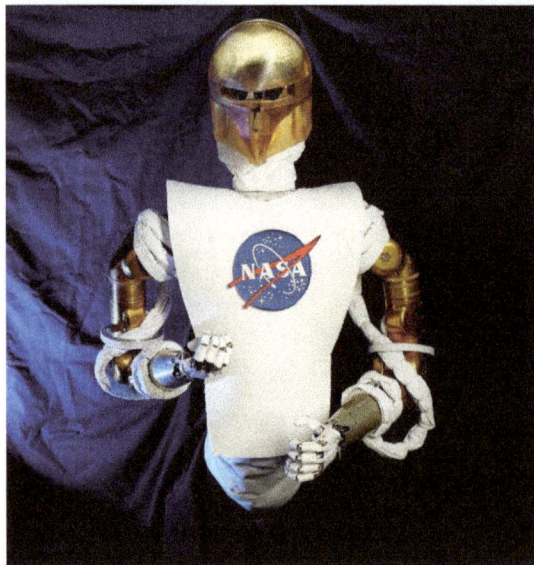

The arms, fingers and head of the JSC Robonaut are modeled as kinematic chains.

Kinematic chain refers to an assembly of rigid bodies connected by joints that is the mathematical model for a mechanical system. As in the familiar use of the word chain, the rigid bodies, or links, are constrained by their connections to other links. An example is the simple open chain formed by links connected in series, like the usual chain, which is the kinematic model for a typical robot manipulator.

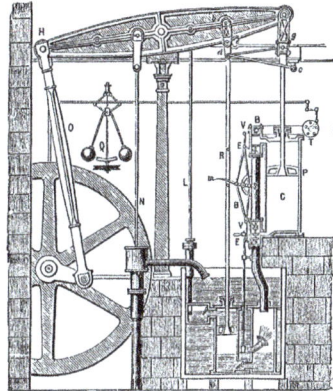

The movement of the Boulton & Watt steam engine is studied as a system of rigid bodies connected by joints forming a kinematic chain.

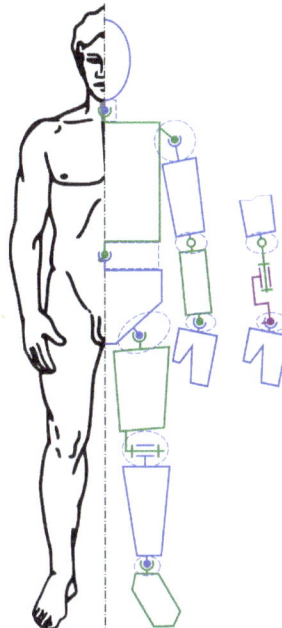

A model of the human skeleton as a kinematic chain allows positioning using forward and inverse kinematics.

Mathematical models of the connections, or joints, between two links are termed kinematic pairs. Kinematic pairs model the hinged and sliding joints fundamental to robotics, often called *lower pairs* and the surface contact joints critical to cams and gearing, called *higher pairs*. These joints are generally modeled as holonomic constraints. A kinematic diagram is a schematic of the mechanical system that shows the kinematic chain.

The modern use of kinematic chains includes compliance that arises from flexure joints in precision mechanisms, link compliance in compliant mechanisms and micro-electro-mechanical systems, and cable compliance in cable robotic and tensegrity systems.

Mobility Formula

The degrees of freedom, or *mobility,* of a kinematic chain is the number of parameters that define the configuration of the chain. A system of n rigid bodies moving in space has $6n$ degrees of freedom measured relative to a fixed frame. This frame is included in the count of bodies, so that mobility does not depend on link that forms the fixed frame. This means the degree-of-freedom of this system is M=6(N-1), where N=n+1 is the number of moving bodies plus the fixed body.

Joints that connect bodies impose constraints. Specifically, hinges and sliders each impose five constraints and therefore remove five degrees of freedom. It is convenient to define the number of constraints c that a joint imposes in terms of the joint's freedom f, where c=6-f. In the case of a hinge or slider, which are one degree of freedom joints, have f=1 and therefore c=6-1=5.

The result is that the mobility of a kinematic chain formed from n moving links and j joints each with freedom f_i, i=1, ..., j, is given by

$$M = 6n - \sum_{i=1}^{j} (6 - f_i) = 6(N - 1 - j) + \sum_{i=1}^{j} f_i$$

Recall that N includes the fixed link.

Analysis of Kinematic Chains

The constraint equations of a kinematic chain couple the range of movement allowed at each joint to the dimensions of the links in the chain, and form algebraic equations that are solved to determine the configuration of the chain associated with specific values of input parameters, called degrees of freedom.

The constraint equations for a kinematic chain are obtained using rigid transformations [Z] to characterize the relative movement allowed at each joint and separate rigid transformations [X] to define the dimensions of each link. In the case of a serial open chain, the result is a sequence of rigid transformations alternating joint and link transformations from the base of the chain to its end link, which is equated to the specified position for the end link. A chain of n links connected in series has the kinematic equations,

$$[T] = [Z_1][X_1][Z_2][X_2]...[X_{n-1}][Z_n],$$

where [T] is the transformation locating the end-link---notice that the chain includes a "zeroth" link consisting of the ground frame to which it is attached. These equations are called the forward kinematics equations of the serial chain.

Kinematic chains of a wide range of complexity are analyzed by equating the kinematics equations of serial chains that form loops within the kinematic chain. These equations are often called *loop equations*.

The complexity (in terms of calculating the forward and inverse kinematics) of the chain is determined by the following factors:

- Its topology: a serial chain, a parallel manipulator, a tree structure, or a graph.

- Its geometrical form: how are neighbouring joints spatially connected to each other?

Explanation:-

Two or more rigid bodies in space are collectively called a rigid body system. We can hinder the motion of these independent rigid bodies with kinematic constraints. Kinematic constraints are constraints between rigid bodies that result in the decrease of the degrees of freedom of rigid body system.

Synthesis of Kinematic Chains

The constraint equations of a kinematic chain can be used in reverse to determine the dimensions of the links from a specification of the desired movement of the system. This is termed *kinematic synthesis*.

Perhaps the most developed formulation of kinematic synthesis is for four-bar linkages, which is known as Burmester theory.

Ferdinand Freudenstein is often called the father of modern kinematics for his contributions to the kinematic synthesis of linkages beginning in the 1950s. His use of the newly developed computer to solve *Freudenstein's equation* became the prototype of computer-aided design systems.

This work has been generalized to the synthesis of spherical and spatial mechanisms.

Rigid Body Dynamics

Rigid-body dynamics studies the movement of systems of interconnected bodies under the action of external forces. The assumption that the bodies are rigid, which means that they do not deform under the action of applied forces, simplifies the analysis by reducing the parameters that describe the configuration of the system to the translation and rotation of reference frames attached to each body. This excludes bodies that display fluid highly elastic, and plastic behavior.

The dynamics of a rigid body system is described by the laws of kinematics and by the application of Newton's second law (kinetics) or their derivative form Lagrangian mechanics. The solution of these equations of motion provides a description of the position, the motion and the acceleration of the individual components of the system and overall the system itself, as a function of time. The formulation and solution of rigid body dynamics is an important tool in the computer simulation of mechanical systems.

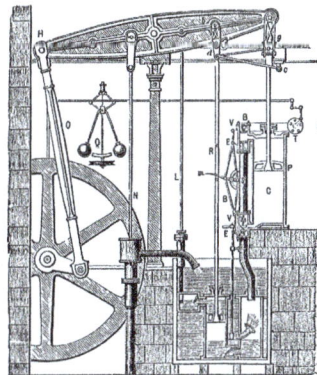

Movement of each of the components of the Boulton & Watt Steam Engine (1784) can be described by a set of equations of kinematics and kinetics

Planar Rigid Body Dynamics

If a system of particles moves parallel to a fixed plane, the system is said to be constrained to planar movement. In this case, Newton's laws (kinetics) for a rigid system of N particles, P_i, i=1,...,N, simplify because there is no movement in the k direction. Determine the resultant force and torque at a reference point R, to obtain

$$\mathbf{F} = \sum_{i=1}^{N} m_i \mathbf{A}_i, \quad \mathbf{T} = \sum_{i=1}^{N} (\mathbf{r}_i - \mathbf{R}) \times (m_i \mathbf{A}_i),$$

where r_i denotes the planar trajectory of each particle.

The kinematics of a rigid body yields the formula for the acceleration of the particle P_i in terms of the position R and acceleration A of the reference particle as well as the angular velocity vector ω and angular acceleration vector α of the rigid system of particles as,

$$\mathbf{A}_i = \alpha \times (\mathbf{r}_i - \mathbf{R}) + \omega \times \omega \times (\mathbf{r}_i - \mathbf{R}) + \mathbf{A}.$$

For systems that are constrained to planar movement, the angular velocity and angular acceleration vectors are directed along k perpendicular to the plane of movement, which simplifies this acceleration equation. In this case, the acceleration vectors can be simplified by introducing the unit vectors e_i from the reference point R to a point r_i and the unit vectors $t_i = k \times e_i$, so

$$\mathbf{A}_i = \alpha(\Delta r_i \mathbf{t}_i) - \omega^2 (\Delta r_i \mathbf{e}_i) + \mathbf{A}.$$

This yields the resultant force on the system as

$$\mathbf{F} = \alpha \sum_{i=1}^{N} m_i (\Delta r_i \mathbf{t}_i) - \omega^2 \sum_{i=1}^{N} m_i (\Delta r_i \mathbf{e}_i) + (\sum_{i=1}^{N} m_i) \mathbf{A},$$

and torque as

$$\mathbf{T} = \sum_{i=1}^{N} (m_i \Delta r_i \mathbf{e}_i) \times (\alpha(\Delta r_i \mathbf{t}_i) - \omega^2 (\Delta r_i \mathbf{e}_i) + \mathbf{A}) = (\sum_{i=1}^{N} m_i \Delta r_i^2) \alpha \vec{k} + (\sum_{i=1}^{N} m_i \Delta r_i \mathbf{e}_i) \times \mathbf{A},$$

where $e_i \times e_i = 0$, and $e_i \times t_i = k$ is the unit vector perpendicular to the plane for all of the particles P_i.

Use the center of mass C as the reference point, so these equations for Newton's laws simplify to become

$$\mathbf{F} = M\mathbf{A}, \quad \mathbf{T} = I_C \alpha \vec{k},$$

where M is the total mass and I_C is the moment of inertia about an axis perpendicular to the movement of the rigid system and through the center of mass.

Rigid Body in Three Dimensions

Orientation or Attitude Descriptions

Several methods to describe orientations of a rigid body in three dimensions have been developed. They are summarized in the following sections.

Euler Angles

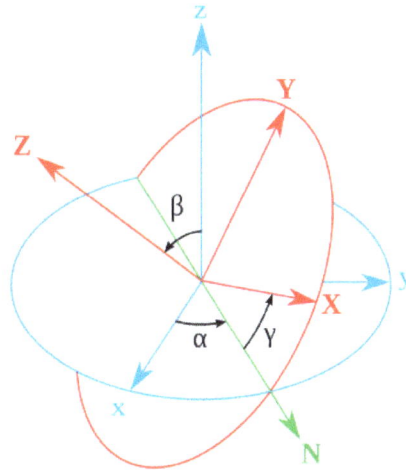

Euler angles, one of the possible ways to describe an orientation.

The first attempt to represent an orientation is attributed to Leonhard Euler. He imagined three reference frames that could rotate one around the other, and realized that by starting with a fixed reference frame and performing three rotations, he could get any other reference frame in the space (using two rotations to fix the vertical axis and other to fix the other two axes). The values of these three rotations are called Euler angles.

Tait–Bryan Angles

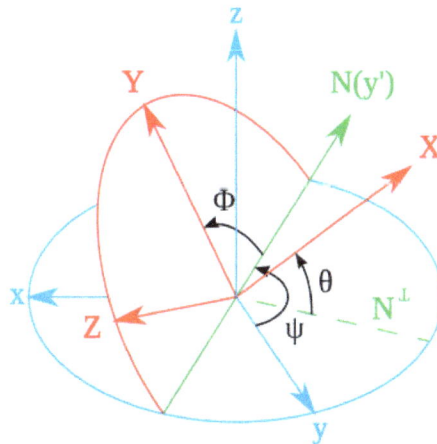

Tait–Bryan angles, another way to describe orientation.

These are three angles, also known as yaw, pitch and roll, Navigation angles and Cardan angles. Mathematically they constitute a set of six possibilities inside the twelve possible sets of Euler angles, the ordering being the one best used for describing the orientation of a vehicle such as an airplane. In aerospace engineering they are usually referred to as Euler angles.

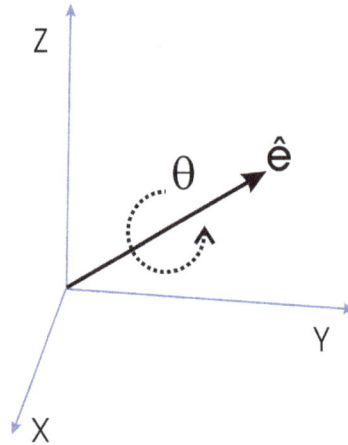

A rotation represented by an Euler axis and angle.

Orientation Vector

Euler also realized that the composition of two rotations is equivalent to a single rotation about a different fixed axis (Euler's rotation theorem). Therefore, the composition of the former three angles has to be equal to only one rotation, whose axis was complicated to calculate until matrices were developed.

Based on this fact he introduced a vectorial way to describe any rotation, with a vector on the rotation axis and module equal to the value of the angle. Therefore, any orientation can be represented by a rotation vector (also called Euler vector) that leads to it from the reference frame. When used to represent an orientation, the rotation vector is commonly called orientation vector, or attitude vector.

A similar method, called axis-angle representation, describes a rotation or orientation using a unit vector aligned with the rotation axis, and a separate value to indicate the angle.

Orientation Matrix

With the introduction of matrices the Euler theorems were rewritten. The rotations were described by orthogonal matrices referred to as rotation matrices or direction cosine matrices. When used to represent an orientation, a rotation matrix is commonly called orientation matrix, or attitude matrix.

The above-mentioned Euler vector is the eigenvector of a rotation matrix (a rotation matrix has a unique real eigenvalue). The product of two rotation matrices is the composition of rotations. Therefore, as before, the orientation can be given as the rotation from the initial frame to achieve the frame that we want to describe.

The configuration space of a non-symmetrical object in n-dimensional space is $SO(n) \times R^n$. Orientation may be visualized by attaching a basis of tangent vectors to an object. The direction in which each vector points determines its orientation.

Orientation Quaternion

Another way to describe rotations is using rotation quaternions, also called versors. They are equivalent to rotation matrices and rotation vectors. With respect to rotation vectors, they can be more easily converted to and from matrices. When used to represent orientations, rotation quaternions are typically called orientation quaternions or attitude quaternions.

Newton's Second Law in Three Dimensions

To consider rigid body dynamics in three-dimensional space, Newton's second law must be extended to define the relationship between the movement of a rigid body and the system of forces and torques that act on it.

Newton's formulated his second law for a particle as, "The change of motion of an object is proportional to the force impressed and is made in the direction of the straight line in which the force is impressed." Because Newton generally referred to mass times velocity as the "motion" of a particle, the phrase "change of motion" refers to the mass times acceleration of the particle, and so this law is usually written as

$$\mathbf{F} = m\mathbf{a},$$

where F is understood to be the only external force acting on the particle, m is the mass of the particle, and a is its acceleration vector. The extension of Newton's second law to rigid bodies is achieved by considering a rigid system of particles.

Rigid System of Particles

If a system of N particles, P_i, i=1,...,N, are assembled into a rigid body, then Newton's second law can be applied to each of the particles in the body. If F_i is the external force applied to particle P_i with mass m_i, then

$$\mathbf{F}_i + \sum \mathbf{F}_{ij} = m_i \mathbf{a}_i, \quad i = 1, \ldots, N,$$

where F_{ij} is the internal force of particle P_j acting on particle P_i that maintains the constant distance between these particles.

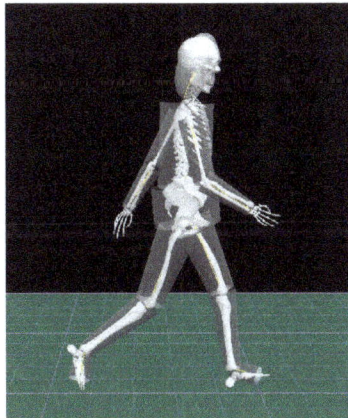

Human body modelled as a system of rigid bodies of geometrical solids. Representative bones were added for better visualization of the walking person.

An important simplification to these force equations is obtained by introducing the resultant force and torque that acts on the rigid system. This resultant force and torque is obtained by choosing one of the particles in the system as a reference point, R, where each of the external forces are applied with the addition of an associated torque. The resultant force F and torque T are given by the formulas,

$$\mathbf{F} = \sum_{i=1}^{N} \mathbf{F}_i, \quad \mathbf{T} = \sum_{i=1}^{N} (\mathbf{R}_i - \mathbf{R}) \times \mathbf{F}_i,$$

where R_i is the vector that defines the position of particle to P_i.

Newton's second law for a particle combines with these formulas for the resultant force and torque to yield,

$$\mathbf{F} = \sum_{i=1}^{N} m_i \mathbf{a}_i, \quad \mathbf{T} = \sum_{i=1}^{N} (\mathbf{R}_i - \mathbf{R}) \times (m_i \mathbf{a}_i),$$

where the internal forces F_{ij} cancel in pairs. The kinematics of a rigid body yields the formula for the acceleration of the particle P_i in terms of the position R and acceleration a of the reference particle as well as the angular velocity vector ω and angular acceleration vector α of the rigid system of particles as,

$$\mathbf{a}_i = \alpha \times (\mathbf{R}_i - \mathbf{R}) + \omega \times (\omega \times (\mathbf{R}_i - \mathbf{R})) + \mathbf{a}.$$

Mass Properties

The mass properties of the rigid body are represented by its center of mass and inertia matrix. Choose the reference point R so that it satisfies the condition

$$\sum_{i=1}^{N} m_i (\mathbf{R}_i - \mathbf{R}) = 0,$$

then it is known as the center of mass of the system. The inertia matrix $[I_R]$ of the system relative to the reference point R is defined by

$$[I_R] = \sum_{i=1}^{N} m_i (\mathbf{I}(\mathbf{S}_i^T \mathbf{S}_i) - \mathbf{S}_i \mathbf{S}_i^T),$$

where S_i is the column vector $R_i - R$; and S_i^T is its transpose.

$S_i^T S_i$ is the scalar product of S_i with itself, while $S_i S_i^T$ is the tensor product of S_i with itself.

I is the 3 by 3 identity matrix.

Force-Torque Equations

Using the center of mass and inertia matrix, the force and torque equations for a single rigid body take the form

$$F = m\mathbf{a}, \quad \mathbf{T} = [I_R]\alpha + \omega \times [I_R]\omega,$$

and are known as Newton's second law of motion for a rigid body.

The dynamics of an interconnected system of rigid bodies, B_j, $j = 1, ..., M$, is formulated by isolating each rigid body and introducing the interaction forces. The resultant of the external and interaction forces on each body, yields the force-torque equations

$$\mathbf{F}_j = m_j \mathbf{a}_j, \quad \mathbf{T}_j = [I_R]_j \alpha_j + \omega_j \times [I_R]_j \omega_j, \quad j = 1, ..., M.$$

Newton's formulation yields $6M$ equations that define the dynamics of a system of M rigid bodies.

Rotation in Three Dimensions

When a rotating object is under the influence of torques, it exhibits the behaviours of precession and nutation. The fundamental equation describing the behavior of a rotating solid body is:

$$\tau = \frac{DL}{Dt} = \frac{dL}{dt} + \omega \times L = \frac{d(I\omega)}{dt} + \omega \times I\omega = I\alpha + \omega \times I\omega$$

where the pseudovectors τ and L are, respectively, the torques on the body and its angular momentum, the scalar I is its moment of inertia, the vector ω is its angular velocity, the vector α is its angular acceleration, D is the differential in an inertial reference frame and d is the differential in a relative reference frame fixed with the body.

It follows from this that a torque τ applied perpendicular to the axis of rotation, and therefore perpendicular to L, results in a rotation about an axis perpendicular to both τ and L. This motion is called *precession*. The angular velocity of precession Ω_p is given by the cross product:

$$\tau = \Omega_p \times L.$$

Precession of a Japanese top (Chikyu Goma).

Precession can be demonstrated by placing a spinning top with its axis horizontal and supported loosely (frictionless toward precession) at one end. Instead of falling, as might be expected, the top appears to defy gravity by remaining with its axis horizontal, when the other end of the axis is left unsupported and the free end of the axis slowly describes a circle in a horizontal plane, the resulting precession turning. This effect is explained by the above equations. The torque on the top is supplied by a couple of forces: gravity acting downward on the device's centre of mass, and an equal force acting upward to support one end of the device. The rotation resulting from this torque is not downward, as might be intuitively expected, causing the device to fall, but perpendicular to both the gravitational torque (horizontal and perpendicular to the axis of rotation) and the axis of rotation (horizontal and outwards from the point of support), i.e., about a vertical axis, causing the device to rotate slowly about the supporting point.

Under a constant torque of magnitude τ, the speed of precession Ω_p is inversely proportional to L, the magnitude of its angular momentum:

$$\tau = \Omega_p L \sin \theta,$$

where θ is the angle between the vectors Ω_p and L. Thus, if the top's spin slows down (for example, due to friction), its angular momentum decreases and so the rate of precession increases. This continues until the device is unable to rotate fast enough to support its own weight, when it stops precessing and falls off its support, mostly because friction against precession cause another precession that goes to cause the fall.

By convention, these three vectors - torque, spin, and precession - are all oriented with respect to each other according to the right-hand rule.

Virtual Work of Forces Acting on a Rigid Body

An alternate formulation of rigid body dynamics that has a number of convenient features is obtained by considering the virtual work of forces acting on a rigid body.

The virtual work of forces acting at various points on a single rigid body can be calculated using the velocities of their point of application and the resultant force and torque. To see this, let the forces F_1, F_2 ... F_n act on the points R_1, R_2 ... R_n in a rigid body.

The trajectories of R_i, $i = 1, ..., n$ are defined by the movement of the rigid body. The velocity of the points R_i along their trajectories are

$$\mathbf{V}_i = \vec{\omega} \times (\mathbf{R}_i - \mathbf{R}) + \mathbf{V},$$

where ω is the angular velocity vector of the body.

Virtual Work

Work is computed from the dot product of each force with the displacement of its point of contact

$$\delta W = \sum_{i=1}^{n} \mathbf{F}_i \cdot \delta \mathbf{r}_i.$$

If the trajectory of a rigid body is defined by a set of generalized coordinates $q_j, j = 1, ..., m$, then the virtual displacements δr_i are given by

$$\delta \mathbf{r}_i = \sum_{j=1}^{m} \frac{\partial \mathbf{r}_i}{\partial q_j} \delta q_j = \sum_{j=1}^{m} \frac{\partial \mathbf{V}_i}{\partial \dot{q}_j} \delta q_j.$$

The virtual work of this system of forces acting on the body in terms of the generalized coordinates becomes

$$\delta W = \mathbf{F}_1 \cdot \left(\sum_{j=1}^{m} \frac{\partial \mathbf{V}_1}{\partial \dot{q}_j} \delta q_j \right) + \ldots + \mathbf{F}_n \cdot \left(\sum_{j=1}^{m} \frac{\partial \mathbf{V}_n}{\partial \dot{q}_j} \delta q_j \right)$$

or collecting the coefficients of δq_j

$$\delta W = \left(\sum_{i=1}^{n} \mathbf{F}_i \cdot \frac{\partial \mathbf{V}_i}{\partial \dot{q}_1} \right) \delta q_1 + \ldots + \left(\sum_{1=1}^{n} \mathbf{F}_i \cdot \frac{\partial \mathbf{V}_i}{\partial \dot{q}_m} \right) \delta q_m.$$

Generalized Forces

For simplicity consider a trajectory of a rigid body that is specified by a single generalized coordinate q, such as a rotation angle, then the formula becomes

$$\delta W = \left(\sum_{i=1}^{n} \mathbf{F}_i \cdot \frac{\partial \mathbf{V}_i}{\partial \dot{q}} \right) \delta q = \left(\sum_{i=1}^{n} \mathbf{F}_i \cdot \frac{\partial (\vec{\omega} \times (\mathbf{R}_i - \mathbf{R}) + \mathbf{V})}{\partial \dot{q}} \right) \delta q.$$

Introduce the resultant force F and torque T so this equation takes the form

$$\delta W = \left(\mathbf{F} \cdot \frac{\partial \mathbf{V}}{\partial \dot{q}} + \mathbf{T} \cdot \frac{\partial \vec{\omega}}{\partial \dot{q}} \right) \delta q.$$

The quantity Q defined by

$$Q = \mathbf{F} \cdot \frac{\partial \mathbf{V}}{\partial \dot{q}} + \mathbf{T} \cdot \frac{\partial \vec{\omega}}{\partial \dot{q}},$$

is known as the generalized force associated with the virtual displacement δq. This formula generalizes to the movement of a rigid body defined by more than one generalized coordinate, that is

$$\delta W = \sum_{j=1}^{m} Q_j \delta q_j,$$

where

$$Q_j = \mathbf{F} \cdot \frac{\partial \mathbf{V}}{\partial \dot{q}_j} + \mathbf{T} \cdot \frac{\partial \vec{\omega}}{\partial \dot{q}_j}, \quad j = 1, \ldots, m.$$

It is useful to note that conservative forces such as gravity and spring forces are derivable from a potential function $V(q_1, \ldots, q_n)$, known as a potential energy. In this case the generalized forces are given by

$$Q_j = -\frac{\partial V}{\partial q_j}, \quad j = 1, \ldots, m.$$

D'Alembert's form of The Principle of Virtual Work

The equations of motion for a mechanical system of rigid bodies can be determined using D'Alembert's form of the principle of virtual work. The principle of virtual work is used to study the static equilibrium of a system of rigid bodies, however by introducing acceleration terms in Newton's laws this approach is generalized to define dynamic equilibrium.

Static Equilibrium

The static equilibrium of a mechanical system rigid bodies is defined by the condition that the virtual work of the applied forces is zero for any virtual displacement of the system. This is known as the *principle of virtual work*. This is equivalent to the requirement that the generalized forces for any virtual displacement are zero, that is $Q_i=0$.

Let a mechanical system be constructed from n rigid bodies, B_i, i=1,...,n, and let the resultant of the applied forces on each body be the force-torque pairs, F_i and T_i, i=1,...,n. Notice that these applied forces do not include the reaction forces where the bodies are connected. Finally, assume that the velocity V_i and angular velocities ω_i, i=,1...,n, for each rigid body, are defined by a single generalized coordinate q. Such a system of rigid bodies is said to have one degree of freedom.

The virtual work of the forces and torques, F_i and T_i, applied to this one degree of freedom system is given by

$$\delta W = \sum_{i=1}^{n}(\mathbf{F}_i \cdot \frac{\partial \mathbf{V}_i}{\partial \dot{q}} + \mathbf{T}_i \cdot \frac{\partial \vec{\omega}_i}{\partial \dot{q}})\delta q = Q\delta q,$$

where

$$Q = \sum_{i=1}^{n}(\mathbf{F}_i \cdot \frac{\partial \mathbf{V}_i}{\partial \dot{q}} + \mathbf{T}_i \cdot \frac{\partial \vec{\omega}_i}{\partial \dot{q}}),$$

is the generalized force acting on this one degree of freedom system.

If the mechanical system is defined by m generalized coordinates, q_j, j=1,...,m, then the system has m degrees of freedom and the virtual work is given by,

$$\delta W = \sum_{j=1}^{m} Q_j \delta q_j,$$

where

$$Q_j = \sum_{i=1}^{n}(\mathbf{F}_i \cdot \frac{\partial \mathbf{V}_i}{\partial \dot{q}_j} + \mathbf{T}_i \cdot \frac{\partial \vec{\omega}_i}{\partial \dot{q}_j}), \quad j = 1,\ldots,m.$$

is the generalized force associated with the generalized coordinate q_j. The principle of virtual work states that static equilibrium occurs when these generalized forces acting on the system are zero, that is

$$Q_j = 0, \quad j = 1,\ldots,m.$$

These m equations define the static equilibrium of the system of rigid bodies.

Generalized Inertia Forces

Consider a single rigid body which moves under the action of a resultant force F and torque T, with one degree of freedom defined by the generalized coordinate q. Assume the reference point for the resultant force and torque is the center of mass of the body, then the generalized inertia force Q* associated with the generalized coordinate q is given by

$$Q^* = -(M\mathbf{A}) \cdot \frac{\partial \mathbf{V}}{\partial \dot{q}} - ([I_R]\alpha + \omega \times [I_R]\omega) \cdot \frac{\partial \vec{\omega}}{\partial \dot{q}}.$$

This inertia force can be computed from the kinetic energy of the rigid body,

$$T = \frac{1}{2} M\mathbf{V} \cdot \mathbf{V} + \frac{1}{2} \vec{\omega} \cdot [I_R]\vec{\omega},$$

by using the formula

$$Q^* = -\left(\frac{d}{dt} \frac{\partial T}{\partial \dot{q}} - \frac{\partial T}{\partial q} \right).$$

A system of n rigid bodies with m generalized coordinates has the kinetic energy

$$T = \sum_{i=1}^{n} (\frac{1}{2} M\mathbf{V}_i \cdot \mathbf{V}_i + \frac{1}{2} \vec{\omega}_i \cdot [I_R]\vec{\omega}_i),$$

which can be used to calculate the m generalized inertia forces

$$Q_j^* = -\left(\frac{d}{dt} \frac{\partial T}{\partial \dot{q}_j} - \frac{\partial T}{\partial q_j} \right), \quad j = 1, \ldots, m.$$

Dynamic Equilibrium

D'Alembert's form of the principle of virtual work states that a system of rigid bodies is in dynamic equilibrium when the virtual work of the sum of the applied forces and the inertial forces is zero for any virtual displacement of the system. Thus, dynamic equilibrium of a system of n rigid bodies with m generalized coordinates requires that

$$\delta W = (Q_1 + Q_1^*)\delta q_1 + \ldots + (Q_m + Q_m^*)\delta q_m = 0,$$

for any set of virtual displacements δq_j. This condition yields m equations,

$$Q_j + Q_j^* = 0, \quad j = 1, \ldots, m,$$

which can also be written as

$$\frac{d}{dt}\frac{\partial T}{\partial \dot{q}_j} - \frac{\partial T}{\partial q_j} = Q_j, \quad j = 1, \ldots, m.$$

The result is a set of m equations of motion that define the dynamics of the rigid body system.

Lagrange's Equations

If the generalized forces Q_j are derivable from a potential energy $V(q_1, \ldots, q_m)$, then these equations of motion take the form

$$\frac{d}{dt}\frac{\partial T}{\partial \dot{q}_j} - \frac{\partial T}{\partial q_j} = -\frac{\partial V}{\partial q_j}, \quad j = 1, \ldots, m.$$

In this case, introduce the Lagrangian, L=T-V, so these equations of motion become

$$\frac{d}{dt}\frac{\partial L}{\partial \dot{q}_j} - \frac{\partial L}{\partial q_j} = 0 \quad j = 1, \ldots, m.$$

These are known as Lagrange's equations of motion.

Linear and Angular Momentum

System of Particles

The linear and angular momentum of a rigid system of particles is formulated by measuring the position and velocity of the particles relative to the center of mass. Let the system of particles P_i, i=1,...,n be located at the coordinates r_i and velocities v_i. Select a reference point R and compute the relative position and velocity vectors,

$$\mathbf{r}_i = (\mathbf{r}_i - \mathbf{R}) + \mathbf{R}, \quad \mathbf{v}_i = \frac{d}{dt}(\mathbf{r}_i - \mathbf{R}) + \mathbf{V}.$$

The total linear and angular momentum vectors relative to the reference point R are

$$\mathbf{p} = \frac{d}{dt}(\sum_{i=1}^{n} m_i(\mathbf{r}_i - \mathbf{R})) + (\sum_{i=1}^{n} m_i)\mathbf{V},$$

and

$$\mathbf{L} = \sum_{i=1}^{n} m_i(\mathbf{r}_i - \mathbf{R}) \times \frac{d}{dt}(\mathbf{r}_i - \mathbf{R}) + (\sum_{i=1}^{n} m_i(\mathbf{r}_i - \mathbf{R})) \times \mathbf{V}.$$

If R is chosen as the center of mass these equations simplify to

$$\mathbf{p} = M\mathbf{V}, \quad \mathbf{L} = \sum_{i=1}^{n} m_i(\mathbf{r}_i - \mathbf{R}) \times \frac{d}{dt}(\mathbf{r}_i - \mathbf{R}).$$

Rigid System of Particles

To specialize these formulas to a rigid body, assume the particles are rigidly connected to each other so P_i, i=1,...,n are located by the coordinates r_i and velocities v_i. Select a reference point R and compute the relative position and velocity vectors,

$$\mathbf{r}_i = (\mathbf{r}_i - \mathbf{R}) + \mathbf{R}, \quad \mathbf{v}_i = \omega \times (\mathbf{r}_i - \mathbf{R}) + \mathbf{V},$$

where ω is the angular velocity of the system.

The linear momentum and angular momentum of this rigid system measured relative to the center of mass R is

$$\mathbf{p} = (\sum_{i=1}^{n} m_i)\mathbf{V}, \quad \mathbf{L} = \sum_{i=1}^{n} m_i(\mathbf{r}_i - \mathbf{R}) \times \mathbf{v}_i = \sum_{i=1}^{n} m_i(\mathbf{r}_i - \mathbf{R}) \times (\omega \times (\mathbf{r}_i - \mathbf{R})).$$

These equations simplify to become,

$$\mathbf{p} = M\mathbf{V}, \quad \mathbf{L} = [I_R]\omega,$$

where M is the total mass of the system and $[I_R]$ is the moment of inertia matrix defined by

$$[I_R] = -\sum_{i=1}^{n} m_i[r_i - R][r_i - R],$$

where $[r_i\text{-}R]$ is the skew-symmetric matrix constructed from the vector $r_i\text{-}R$.

Equations of Motion

In mathematical physics, equations of motion are equations that describe the behaviour of a physical system in terms of its motion as a function of time. More specifically, the equations of motion describe the behaviour of a physical system as a set of mathematical functions in terms of dynamic variables: normally spatial coordinates and time are used, but others are also possible, such as momentum components and time. The most general choice are generalized coordinates which can be any convenient variables characteristic of the physical system. The functions are defined in a Euclidean space in classical mechanics, but are replaced by curved spaces in relativity. If the dynamics of a system is known, the equations are the solutions to the differential equations describing the motion of the dynamics.

There are two main descriptions of motion: dynamics and kinematics. Dynamics is general, since momenta, forces and energy of the particles are taken into account. In this instance, sometimes the term refers to the differential equations that the system satisfies (e.g., Newton's second law or Euler–Lagrange equations), and sometimes to the solutions to those equations.

However, kinematics is simpler as it concerns only variables derived from the positions of objects, and time. In circumstances of constant acceleration, these simpler equations of motion are usually referred to as the "SUVAT" equations, arising from the definitions of kinematic quantities: displacement (s), initial velocity (u), final velocity (v), acceleration (a), and time (t).

Equations of motion can therefore be grouped under these main classifiers of motion. In all cases, the main types of motion are translations, rotations, oscillations, or any combinations of these.

A differential equation of motion, usually identified as some physical law and applying definitions of physical quantities, is used to set up an equation for the problem. Solving the differential equation will lead to a general solution with arbitrary constants, the arbitrariness corresponding to a family of solutions. A particular solution can be obtained by setting the initial values, which fixes the values of the constants.

To state this formally, in general an equation of motion M is a function of the position r of the object, its velocity (the first time derivative of r, $v = dr/dt$), and its acceleration (the second derivative of r, $a = d^2r/dt^2$), and time t. Euclidean vectors in 3D are denoted throughout in bold. This is equivalent to saying an equation of motion in r is a second order ordinary differential equation (ODE) in r,

$$M\left[\mathbf{r}(t), \dot{\mathbf{r}}(t), \ddot{\mathbf{r}}(t), t\right] = 0,$$

where t is time, and each overdot denotes one time derivative. The initial conditions are given by the *constant* values at $t = 0$,

$$\mathbf{r}(0), \quad \dot{\mathbf{r}}(0).$$

The solution r(t) to the equation of motion, with specified initial values, describes the system for all times t after $t = 0$. Other dynamical variables like the momentum p of the object, or quantities derived from r and p like angular momentum, can be used in place of r as the quantity to solve for from some equation of motion, although the position of the object at time t is by far the most sought-after quantity.

Sometimes, the equation will be linear and is more likely to be exactly solvable. In general, the equation will be non-linear, and cannot be solved exactly so a variety of approximations must be used. The solutions to nonlinear equations may show chaotic behavior depending on how *sensitive* the system is to the initial conditions.

History

Historically, equations of motion first appeared in classical mechanics to describe the motion of massive objects, a notable application was to celestial mechanics to predict the motion of the planets as if they orbit like clockwork (this was how Neptune was predicted before its discovery), and also investigate the stability of the solar system.

It is important to observe that the huge body of work involving kinematics, dynamics and the mathematical models of the universe developed in baby steps – faltering, getting up and correcting itself – over three millennia and included contributions of both known names and others who have since faded from the annals of history.

In antiquity, notwithstanding the success of priests, astrologers and astronomers in predicting solar and lunar eclipses, the solstices and the equinoxes of the Sun and the period of the Moon, there was nothing other than a set of algorithms to help them. Despite the great strides made in the development of geometry made by Ancient Greeks and surveys in Rome, we were to wait for another thousand years before the first equations of motion arrive.

The exposure of Europe to the collected works by the Muslims of the Greeks, the Indians and the Islamic scholars, such as Euclid's *Elements*, the works of Archimedes, and Al-Khwārizmī's treatises began in Spain, and scholars from all over Europe went to Spain, read, copied and translated the learning into Latin. The exposure of Europe to Arabic numerals and their ease in computations encouraged first the scholars to learn them and then the merchants and invigorated the spread of knowledge throughout Europe.

By the 13th century the universities of Oxford and Paris had come up, and the scholars were now studying mathematics and philosophy with lesser worries about mundane chores of life—the fields were not as clearly demarcated as they are in the modern times. Of these, compendia and redactions, such as those of Johannes Campanus, of Euclid and Aristotle, confronted scholars with ideas about infinity and the ratio theory of elements as a means of expressing relations between various quantities involved with moving bodies. These studies led to a new body of knowledge that is now known as physics.

Of these institutes Merton College sheltered a group of scholars devoted to natural science, mainly physics, astronomy and mathematics, of similar in stature to the intellectuals at the University of Paris. Thomas Bradwardine, one of those scholars, extended Aristotelian quantities such as distance and velocity, and assigned intensity and extension to them. Bradwardine suggested an exponential law involving force, resistance, distance, velocity and time. Nicholas Oresme further extended Bradwardine's arguments. The Merton school proved that the quantity of motion of a body undergoing a uniformly accelerated motion is equal to the quantity of a uniform motion at the speed achieved halfway through the accelerated motion.

For writers on kinematics before Galileo, since small time intervals could not be measured, the affinity between time and motion was obscure. They used time as a function of distance, and in free fall, greater velocity as a result of greater elevation. Only Domingo de Soto, a Spanish theologian, in his commentary on Aristotle's *Physics* published in 1545, after defining "uniform difform" motion (which is uniformly accelerated motion) – the word velocity wasn't used – as proportional to time, declared correctly that this kind of motion was identifiable with freely falling bodies and projectiles, without his proving these propositions or suggesting a formula relating time, velocity and distance. De Soto's comments are shockingly correct regarding the definitions of acceleration (acceleration was a rate of change of motion (velocity) in time) and the observation that during the violent motion of ascent acceleration would be negative.

Discourses such as these spread throughout Europe and definitely influenced Galileo and others, and helped in laying the foundation of kinematics. Galileo deduced the equation $s = 1/2gt^2$ in his work geometrically, using Merton's rule, now known as a special case of one of the equations of kinematics. He couldn't use the now-familiar mathematical reasoning. The relationships between speed, distance, time and acceleration was not known at the time.

Galileo was the first to show that the path of a projectile is a parabola. Galileo had an understanding of centrifugal force and gave a correct definition of momentum. This emphasis of momentum as a fundamental quantity in dynamics is of prime importance. He measured momentum by the product of velocity and weight; mass is a later concept, developed by Huygens and Newton. In the swinging of a simple pendulum, Galileo says in *Discourses* that "every momentum acquired in the descent along an arc is equal to that which causes the same moving body to ascend through the same arc." His analysis on projectiles indicates that Galileo had grasped the first law and the second law of motion. He did not generalize and make them applicable to bodies not subject to the earth's gravitation. That step was Newton's contribution.

The term "inertia" was used by Kepler who applied it to bodies at rest.The first law of motion is now often called the law of inertia.

Galileo did not fully grasp the third law of motion, the law of the equality of action and reaction, though he corrected some errors of Aristotle. With Stevin and others Galileo also wrote on statics. He formulated the principle of the parallelogram of forces, but he did not fully recognize its scope.

Galileo also was interested by the laws of the pendulum, his first observations was when he was a young man. In 1583, while he was praying in the cathedral at Pisa, his attention was arrested by the motion of the great lamp lighted and left swinging, referencing his own pulse for time keeping. To him the period appeared the same, even after the motion had greatly diminished, discovering the isochronism of the pendulum.

More careful experiments carried out by him later, and described in his Discourses, revealed the period of oscillation varies with the square root of length but is independent of the mass the pendulum.

Thus we arrive at René Descartes, Isaac Newton, Gottfried Leibniz, et al.; and the evolved forms of the equations of motion that begin to be recognized as the modern ones.

Later the equations of motion also appeared in electrodynamics, when describing the motion of charged particles in electric and magnetic fields, the Lorentz force is the general equation which serves as the definition of what is meant by an electric field and magnetic field. With the advent of special relativity and general relativity, the theoretical modifications to spacetime meant the classical equations of motion were also modified to account for the finite speed of light, and curvature of spacetime. In all these cases the differential equations were in terms of a function describing the particle's trajectory in terms of space and time coordinates, as influenced by forces or energy transformations.

However, the equations of quantum mechanics can also be considered "equations of motion", since they are differential equations of the wavefunction, which describes how a quantum state behaves analogously using the space and time coordinates of the particles. There are analogs of equations of motion in other areas of physics, for collections of physical phenomena that can be considered waves, fluids, or fields.

Kinematic Equations for One Particle

Kinematic Quantities

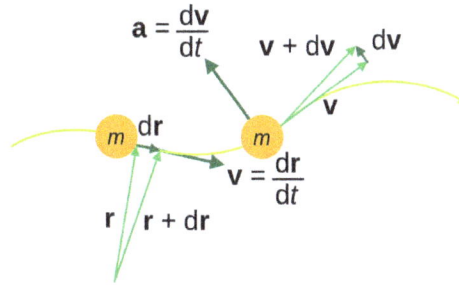

Kinematic quantities of a classical particle of mass m: position r, velocity v, acceleration a.

From the instantaneous position r = r(t), instantaneous meaning at an instant value of time t, the instantaneous velocity v = v(t) and acceleration a = a(t) have the general, coordinate-independent definitions;

$$\mathbf{v} = \frac{d\mathbf{r}}{dt}, \quad \mathbf{a} = \frac{d\mathbf{v}}{dt} = \frac{d^2\mathbf{r}}{dt^2},$$

Notice that velocity always points in the direction of motion, in other words for a curved path it is the tangent vector. Loosely speaking, first order derivatives are related to tangents of curves. Still for curved paths, the acceleration is directed towards the center of curvature of the path. Again, loosely speaking, second order derivatives are related to curvature.

The rotational analogues are the "angular vector" (angle the particle rotates about some axis) $\theta = \theta(t)$, angular velocity $\omega = \omega(t)$, and angular acceleration $\alpha = \alpha(t)$:

$$\theta = \theta\hat{n}, \quad \omega = \frac{d\theta}{dt}, \quad \alpha = \frac{d\omega}{dt},$$

where n̂ is a unit vector in the direction of the axis of rotation, and θ is the angle the object turns through about the axis.

The following relation holds for a point-like particle, orbiting about some axis with angular velocity ω:

$$v = \omega \times r$$

where r is the position vector of the particle (radial from the rotation axis) and v the tangential velocity of the particle. For a rotating continuum rigid body, these relations hold for each point in the rigid body.

Uniform Acceleration

The differential equation of motion for a particle of constant or uniform acceleration in a straight line is simple: the acceleration is constant, so the second derivative of the position of the object is constant. The results of this case are summarized below.

Constant Translational Acceleration in a Straight Line

These equations apply to a particle moving linearly, in three dimensions in a straight line with constant acceleration. Since the position, velocity, and acceleration are collinear (parallel, and lie on the same line) – only the magnitudes of these vectors are necessary, and because the motion is along a straight line, the problem effectively reduces from three dimensions to one.

$$v = at + v_0 \quad [1]$$

$$r = r_0 + v_0 t + \tfrac{1}{2} at^2 \quad [2]$$

$$r = r_0 + \tfrac{1}{2}(v + v_0)t \quad [3]$$

$$v^2 = v_0^2 + 2a(r - r_0) \quad [4]$$

$$r = r_0 + vt - \tfrac{1}{2}at^2 \quad [5]$$

where:

- r_0 is the particle's initial position
- r is the particle's final position
- v_0 is the particle's initial velocity
- v is the particle's final velocity
- a is the particle's acceleration
- t is the time interval

Derivation

Equations and are from integrating the definitions of velocity and acceleration, subject to the initial conditions $r(t_o) = r_o$ and $v(t_o) = v_o$;

$$\mathbf{v} = \int \mathbf{a}dt = \mathbf{a}t + \mathbf{v}_0, \quad [1] \quad , \quad \mathbf{r} = \int (\mathbf{a}t + \mathbf{v}_0)dt = \frac{\mathbf{a}t^2}{2} + \mathbf{v}_0 t + \mathbf{r}_0, \quad [2]$$

in magnitudes,

$$v = at + v_0, \quad [1] \quad , \quad r = \frac{at^2}{2} + v_0 t + r_0. \quad [2]$$

Equation involves the average velocity $v + v_{o/2}$. Intuitively, the velocity increases linearly, so the average velocity multiplied by time is the distance traveled while increasing the velocity from v_o to v, as can be illustrated graphically by plotting velocity against time as a straight line graph. Algebraically, it follows from solving for

$$\mathbf{a} = \frac{(\mathbf{v} - \mathbf{v}_0)}{t}$$

and substituting into

$$\mathbf{r} = \mathbf{r}_0 + \mathbf{v}_0 t + \frac{t}{2}(\mathbf{v} - \mathbf{v}_0),$$

then simplifying to get

$$\mathbf{r} = \mathbf{r}_0 + \frac{t}{2}(\mathbf{v} + \mathbf{v}_0)$$

or in magnitudes

$$r = r_0 + \left(\frac{v + v_0}{2}\right)t \quad [3]$$

From [3],

$$t = (r - r_0)\left(\frac{2}{v + v_0}\right)$$

substituting for t in :

$$v = a(r - r_0)\left(\frac{2}{v + v_0}\right) + v_0$$
$$v(v + v_0) = 2a(r - r_0) + v_0(v + v_0)$$
$$v^2 + vv_0 = 2a(r - r_0) + v_0 v + v_0^2$$
$$v^2 = v_0^2 + 2a(r - r_0) \quad [4]$$

From [3],

$$2(r - r_0) - vt = v_0 t$$

substituting into [2]:

$$r = \frac{at^2}{2} + 2r - 2r_0 - vt + r_0$$
$$0 = \frac{at^2}{2} + r - r_0 - vt$$
$$r = r_0 + vt - \frac{at^2}{2} \quad [5]$$

Usually only the first 4 are needed, the fifth is optional.

Here a is *constant* acceleration, or in the case of bodies moving under the influence of gravity, the standard gravity g is used. Note that each of the equations contains four of the five variables, so in this situation it is sufficient to know three out of the five variables to calculate the remaining two.

In elementary physics the same formulae are frequently written in different notation as:

$$dv = u + at \quad [1]$$
$$s = ut + \tfrac{1}{2}at^2 \quad [2]$$
$$s = \tfrac{1}{2}(u+v)t \quad [3]$$
$$v^2 = u^2 + 2as \quad [4]$$
$$s = vt - \tfrac{1}{2}at^2 \quad [5]$$

where u has replaced v_0, s replaces r, and $s_0 = 0$. They are often referred to as the "SUVAT" equations, where "SUVAT" is an acronym from the variables: s = displacement (s_0 = initial displacement), u = initial velocity, v = final velocity, a = acceleration, t = time.

Constant Linear Acceleration in Any Direction

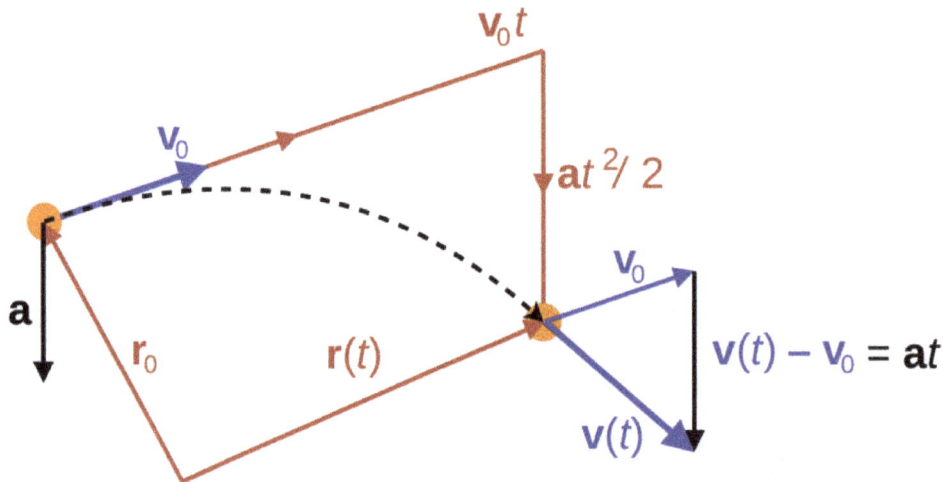

Trajectory of a particle with initial position vector $\mathbf{r}_0$ and velocity $\mathbf{v}_0$, subject to constant acceleration $\mathbf{a}$, all three quantities in any direction, and the position $\mathbf{r}(t)$ and velocity $\mathbf{v}(t)$ after time t.

The initial position, initial velocity, and acceleration vectors need not be collinear, and take an almost identical form. The only difference is that the square magnitudes of the velocities require the dot product. The derivations are essentially the same as in the collinear case,

$$\mathbf{v} = \mathbf{a}t + \mathbf{v}_0 \quad [1]$$
$$\mathbf{r} = \mathbf{r}_0 + \mathbf{v}_0 t + \tfrac{1}{2}\mathbf{a}t^2 \quad [2]$$
$$\mathbf{r} = \mathbf{r}_0 + \tfrac{1}{2}(\mathbf{v}+\mathbf{v}_0)t \quad [3]$$
$$v^2 = v_0^2 + 2\mathbf{a}\cdot(\mathbf{r}-\mathbf{r}_0) \quad [4]$$
$$\mathbf{r} = \mathbf{r}_0 + \mathbf{v}t - \tfrac{1}{2}\mathbf{a}t^2 \quad [5]$$

although the Torricelli equation can be derived using the distributive property of the dot product as follows:

$$v^2 = \mathbf{v}\cdot\mathbf{v} = (\mathbf{v}_0 + \mathbf{a}t)\cdot(\mathbf{v}_0 + \mathbf{a}t) = v_0^2 + 2t(\mathbf{a}\cdot\mathbf{v}_0) + a^2 t^2$$

$$(2\mathbf{a}) \cdot (\mathbf{r} - \mathbf{r}_0) = (2\mathbf{a}) \cdot \left(\mathbf{v}_0 t + \tfrac{1}{2}\mathbf{a}t^2\right) = 2t(\mathbf{a} \cdot \mathbf{v}_0) + a^2 t^2 = v^2 - v_0^2$$

$$\therefore v^2 = v_0^2 + 2(\mathbf{a} \cdot (\mathbf{r} - \mathbf{r}_0))$$

Applications

Elementary and frequent examples in kinematics involve projectiles, for example a ball thrown upwards into the air. Given initial speed u, one can calculate how high the ball will travel before it begins to fall. The acceleration is local acceleration of gravity g. At this point one must remember that while these quantities appear to be scalars, the direction of displacement, speed and acceleration is important. They could in fact be considered as unidirectional vectors. Choosing s to measure up from the ground, the acceleration a must be in fact $-g$, since the force of gravity acts downwards and therefore also the acceleration on the ball due to it.

At the highest point, the ball will be at rest: therefore v = 0. Using equation in the set above, we have:

$$s = \frac{v^2 - u^2}{-2g}.$$

Substituting and cancelling minus signs gives:

$$s = \frac{u^2}{2g}.$$

Constant Circular Acceleration

The analogues of the above equations can be written for rotation. Again these axial vectors must all be parallel to the axis of rotation, so only the magnitudes of the vectors are necessary,

$$\omega = \omega_0 + \alpha t$$

$$\theta = \theta_0 + \omega_0 t + \tfrac{1}{2}\alpha t^2$$

$$\theta = \theta_0 + \tfrac{1}{2}(\omega_0 + \omega)t$$

$$\omega^2 = \omega_0^2 + 2\alpha(\theta - \theta_0)$$

$$\theta = \theta_0 + \omega t - \tfrac{1}{2}\alpha t^2$$

where α is the constant angular acceleration, ω is the angular velocity, ω_0 is the initial angular velocity, θ is the angle turned through (angular displacement), θ_0 is the initial angle, and t is the time taken to rotate from the initial state to the final state.

General Planar Motion

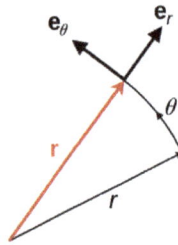

Position vector r, always points radially from the origin.

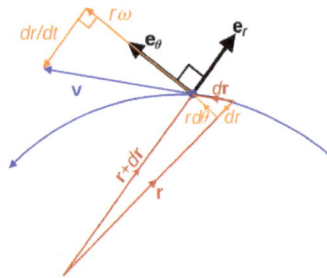

Velocity vector v, always tangent to the path of motion.

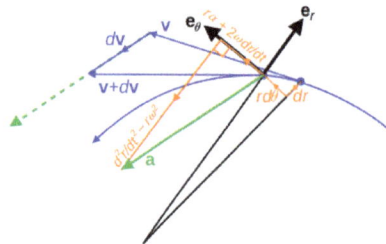

Acceleration vector a, not parallel to the radial motion but offset by the angular and Coriolis accel-erations, nor tangent to the path but offset by the centripetal and radial accelerations.

Kinematic vectors in plane polar coordinates. Notice the setup is not restricted to 2D space, but a plane in any higher dimension.

These are the kinematic equations for a particle traversing a path in a plane, described by position $r = r(t)$. They are simply the time derivatives of the position vector in plane polar coordinates using the definitions of physical quantities above for angular velocity ω and angular acceleration α.

The position, velocity and acceleration of the particle are respectively:

$$\mathbf{r} = \mathbf{r}\left(r(t), \theta(t)\right) = r\hat{\mathbf{e}}_r$$

$$\mathbf{v} = \hat{\mathbf{e}}_r \frac{dr}{dt} + r\omega\hat{\mathbf{e}}_\theta$$

$$\mathbf{a} = \left(\frac{d^2 r}{dt^2} - r\omega^2\right)\hat{\mathbf{e}}_r + \left(r\alpha + 2\omega\frac{dr}{dt}\right)\hat{\mathbf{e}}_\theta$$

where $\hat{e}_r$ and $\hat{e}_\theta$ are the polar unit vectors. For the velocity v, dr/dt is the component of velocity in the radial direction, and $r\omega$ is the additional component due to the rotation. For the acceleration a, $-r\omega^2$ is the centripetal acceleration and $2\omega dr/dt$ the Coriolis acceleration, in addition to the radial acceleration d^2r/dt^2 and angular acceleration $r\alpha$.

Special cases of motion described be these equations are summarized qualitatively in the table below. Two have already been discussed above, in the cases that either the radial components or the angular components are zero, and the non-zero component of motion describes uniform acceleration.

State of motion	Constant r	r linear in t	r quadratic in t	r non-linear in t
Constant θ	Stationary	Uniform translation (constant translational velocity)	Uniform translational acceleration	Non-uniform translation
θ linear in t	Uniform angular motion in a circle (constant angular velocity)	Uniform angular motion in a spiral, constant radial velocity	Angular motion in a spiral, constant radial acceleration	Angular motion in a spiral, varying radial acceleration
θ quadratic in t	Uniform angular acceleration in a circle	Uniform angular acceleration in a spiral, constant radial velocity	Uniform angular acceleration in a spiral, constant radial acceleration	Uniform angular acceleration in a spiral, varying radial acceleration
θ non-linear in t	Non-uniform angular acceleration in a circle	Non-uniform angular acceleration in a spiral, constant radial velocity	Non-uniform angular acceleration in a spiral, constant radial acceleration	Non-uniform angular acceleration in a spiral, varying radial acceleration

General 3D Motion

In 3D space, the equations in spherical coordinates (r, θ, φ) with corresponding unit vectors $\hat{e}_r$, $\hat{e}_\theta$ and $\hat{e}_\varphi$, the position, velocity, and acceleration generalize respectively to

$$\mathbf{r} = \mathbf{r}(t) = r\hat{e}_r$$

$$\mathbf{v} = v\hat{e}_r + r\frac{d\theta}{dt}\hat{e}_\theta + r\frac{d\varphi}{dt}\sin\theta\hat{e}_\varphi$$

$$\mathbf{a} = \left(a - r\left(\frac{d\theta}{dt}\right)^2 - r\left(\frac{d\varphi}{dt}\right)^2\sin^2\theta\right)\hat{e}_r + \left(r\frac{d^2\theta}{dt^2} + 2v\frac{d\theta}{dt} - r\left(\frac{d\varphi}{dt}\right)^2\sin\theta\cos\theta\right)\hat{e}_\theta + \left(r\frac{d^2\varphi}{dt^2}\sin\theta + 2v\frac{d\varphi}{dt}\sin\theta + 2r\frac{d\theta}{dt}\frac{d\varphi}{dt}\cos\theta\right)\hat{e}_\varphi$$

In the case of a constant φ this reduces to the planar equations above.

Dynamic Equations of Motion

Newtonian Mechanics

The first general equation of motion developed was Newton's second law of motion, in its most general form states the rate of change of momentum p = p(t) = mv(t) of an object equals the force

$F = F(x(t), v(t), t)$ acting on it,

$$F = \frac{d\mathbf{p}}{dt}$$

The force in the equation is *not* the force the object exerts. Replacing momentum by mass times velocity, the law is also written more famously as

$$\mathbf{F} = m\mathbf{a}$$

since m is a constant in Newtonian mechanics.

Newton's second law applies to point-like particles, and to all points in a rigid body. They also apply to each point in a mass continua, like deformable solids or fluids, but the motion of the system must be accounted for. In the case the mass is not constant, it is not sufficient to use the product rule for the time derivative on the mass and velocity, and Newton's second law requires some modification consistent with conservation of momentum.

It may be simple to write down the equations of motion in vector form using Newton's laws of motion, but the components may vary in complicated ways with spatial coordinates and time, and solving them is not easy. Often there is an excess of variables to solve for the problem completely, so Newton's laws are not always the most efficient way to determine the motion of a system. In simple cases of rectangular geometry, Newton's laws work fine in Cartesian coordinates, but in other coordinate systems can become dramatically complex.

The momentum form is preferable since this is readily generalized to more complex systems, generalizes to special and general relativity. It can also be used with the momentum conservation. However, Newton's laws are not more fundamental than momentum conservation, because Newton's laws are merely consistent with the fact that zero resultant force acting on an object implies constant momentum, while a resultant force implies the momentum is not constant. Momentum conservation is always true for an isolated system not subject to resultant forces.

For a number of particles, the equation of motion for one particle i influenced by other particles is

$$\frac{d\mathbf{p}_i}{dt} = \mathbf{F}_E + \sum_{i \neq j} \mathbf{F}_{ij}$$

where p_i is the momentum of particle i, F_{ij} is the force on particle i by particle j, and F_E is the resultant external force due to any agent not part of system. Particle i does not exert a force on itself.

Euler's laws of motion are similar to Newton's laws, but they are applied specifically to the motion of rigid bodies. The Newton–Euler equations combine the forces and torques acting on a rigid body into a single equation.

Newton's second law for rotation takes a similar form to the translational case,

$$\tau = \frac{d\mathbf{L}}{dt},$$

by equating the torque acting on the body to the rate of change of its angular momentum L. Analogous to mass times acceleration, the moment of inertia tensor I depends on the distribution of mass about the axis of rotation, and the angular acceleration is the rate of change of angular velocity,

$$\tau = \mathbf{I} \cdot \alpha.$$

Again, these equations apply to point like particles, or at each point of a rigid body.

Likewise, for a number of particles, the equation of motion for one particle i is

$$\frac{d\mathbf{L}_i}{dt} = \tau_E + \sum_{i \neq j} \tau_{ij},$$

where L_i is the angular momentum of particle i, τ_{ij} the torque on particle i by particle j, and τ_E is resultant external torque (due to any agent not part of system). Particle i does not exert a torque on itself.

Applications

Some examples of Newton's law include describing the motion of a simple pendulum,

$$-mg \sin \theta = m \frac{d^2(l\theta)}{dt^2} \quad \Rightarrow \quad \frac{d^2\theta}{dt^2} = -\frac{g}{l} \sin \theta,$$

and a damped, sinusoidally driven harmonic oscillator,

$$F_0 \sin(\omega t) = m \left(\frac{d^2 x}{dt^2} + 2\zeta\omega_0 \frac{dx}{dt} + \omega_0^2 x \right).$$

For describing the motion of masses due to gravity, Newton's law of gravity can be combined with Newton's second law. For two examples, a ball of mass m thrown in the air, in air currents (such as wind) described by a vector field of resistive forces R = R(r, t),

$$-\frac{GmM}{|\mathbf{r}|^2} \hat{\mathbf{e}}_r + \mathbf{R} = m \frac{d^2 \mathbf{r}}{dt^2} + 0 \quad \Rightarrow \quad \frac{d^2 \mathbf{r}}{dt^2} = -\frac{GM}{|\mathbf{r}|^2} \hat{\mathbf{e}}_r + \mathbf{A}$$

where G is the gravitational constant, M the mass of the Earth, and A = R/m is the acceleration of the projectile due to the air currents at position r and time t.

The classical N-body problem for N particles each interacting with each other due to gravity is a set of N nonlinear coupled second order ODEs,

$$\frac{d^2 \mathbf{r}_i}{dt^2} = G \sum_{i \neq j} \frac{m_i m_j}{|\mathbf{r}_j - \mathbf{r}_i|^3} (\mathbf{r}_j - \mathbf{r}_i)$$

where i = 1, 2, ..., N labels the quantities (mass, position, etc.) associated with each particle.

Analytical Mechanics

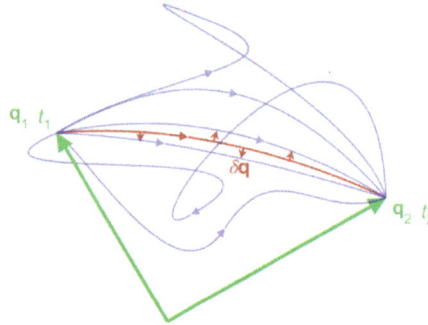

As the system evolves, **q** traces a path through configuration space (only some are shown). The path taken by the system (red) has a stationary action ($\delta S = 0$) under small changes in the configuration of the system ($\delta \mathbf{q}$).

Using all three coordinates of 3D space is unnecessary if there are constraints on the system. If the system has N degrees of freedom, then one can use a set of N generalized coordinates $q(t) = [q_1(t), q_2(t) \dots q_N(t)]$, to define the configuration of the system. They can be in the form of arc lengths or angles. They are a considerable simplification to describe motion, since they take advantage of the intrinsic constraints that limit the system's motion, and the number of coordinates is reduced to a minimum. The time derivatives of the generalized coordinates are the *generalized velocities*

$$\dot{\mathbf{q}} = \frac{d\mathbf{q}}{dt}.$$

The Euler–Lagrange equations are

$$\frac{d}{dt}\left(\frac{\partial L}{\partial \dot{\mathbf{q}}}\right) = \frac{\partial L}{\partial \mathbf{q}},$$

where the *Lagrangian* is a function of the configuration q and its time rate of change dq/dt (and possibly time t)

$$L = L\left[\mathbf{q}(t), \dot{\mathbf{q}}(t), t\right].$$

Setting up the Lagrangian of the system, then substituting into the equations and evaluating the partial derivatives and simplifying, a set of coupled N second order ODEs in the coordinates are obtained.

Hamilton's equations are

$$\dot{\mathbf{p}} = -\frac{\partial H}{\partial \mathbf{q}}, \quad \dot{\mathbf{q}} = +\frac{\partial H}{\partial \mathbf{p}},$$

where the Hamiltonian

$$H = H\left[\mathbf{q}(t), \mathbf{p}(t), t\right],$$

is a function of the configuration q and conjugate *"generalized" momenta*

$$\mathbf{p} = \frac{\partial L}{\partial \dot{\mathbf{q}}},$$

in which $\partial / \partial q = (\partial / \partial q_1, \partial / \partial q_2, ..., \partial / \partial q_N)$ is a shorthand notation for a vector of partial derivatives with respect to the indicated variables, and possibly time t,

Setting up the Hamiltonian of the system, then substituting into the equations and evaluating the partial derivatives and simplifying, a set of coupled $2N$ first order ODEs in the coordinates q_i and momenta p_i are obtained.

The Hamilton–Jacobi equation is

$$-\frac{\partial S(\mathbf{q},t)}{\partial t} = H\left(\mathbf{q},\mathbf{p},t\right).$$

where

$$S[\mathbf{q},t] = \int_{t_1}^{t_2} L(\mathbf{q},\dot{\mathbf{q}},t)\,dt,$$

is *Hamilton's principal function*, also called the *classical action* is a functional of L. In this case, the momenta are given by

$$\mathbf{p} = \frac{\partial S}{\partial \mathbf{q}}.$$

Although the equation has a simple general form, for a given Hamiltonian it is actually a single first order *non-linear* PDE, in $N + 1$ variables. The action S allows identification of conserved quantities for mechanical systems, even when the mechanical problem itself cannot be solved fully, because any differentiable symmetry of the action of a physical system has a corresponding conservation law, a theorem due to Emmy Noether.

All classical equations of motion can be derived from the variational principle known as Hamilton's principle of least action

$$\delta S = 0,$$

stating the path the system takes through the configuration space is the one with the least action S.

Electrodynamics

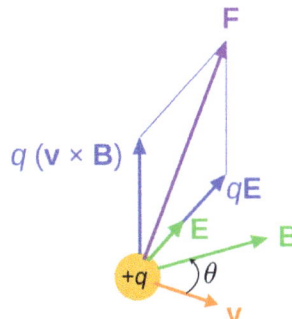

Lorentz force f on a charged particle (of charge q) in motion (instantaneous velocity v). The E field and B field vary in space and time.

In electrodynamics, the force on a charged particle of charge q is the Lorentz force:

$$\mathbf{F} = q\left(\mathbf{E} + \mathbf{v} \times \mathbf{B}\right)$$

Combining with Newton's second law gives a first order differential equation of motion, in terms of position of the particle:

$$m\frac{d^2\mathbf{r}}{dt^2} = q\left(\mathbf{E} + \frac{d\mathbf{r}}{dt} \times \mathbf{B}\right)$$

or its momentum:

$$\frac{d\mathbf{p}}{dt} = q\left(\mathbf{E} + \frac{\mathbf{p} \times \mathbf{B}}{m}\right)$$

The same equation can be obtained using the Lagrangian (and applying Lagrange's equations above) for a charged particle of mass m and charge q:

$$L = \frac{1}{2}m\dot{\mathbf{r}} \cdot \dot{\mathbf{r}} + q\mathbf{A} \cdot \dot{\mathbf{r}} - q\phi$$

where A and ϕ are the electromagnetic scalar and vector potential fields. The Lagrangian indicates an additional detail: the canonical momentum in Lagrangian mechanics is given by:

$$\mathbf{P} = \frac{\partial L}{\partial \dot{\mathbf{r}}} = m\dot{\mathbf{r}} + q\mathbf{A}$$

instead of just mv, implying the motion of a charged particle is fundamentally determined by the mass and charge of the particle. The Lagrangian expression was first used to derive the force equation.

Alternatively the Hamiltonian (and substituting into the equations):

$$H = \frac{\left(\mathbf{P} - q\mathbf{A}\right)^2}{2m} + q\phi$$

can derive the Lorentz force equation.

General Relativity

Geodesic Equation of Motion

The above equations are valid in flat spacetime. In curved space spacetime, things become mathematically more complicated since there is no straight line; this is generalized and replaced by a *geodesic* of the curved spacetime (the shortest length of curve between two points). For curved manifolds with a metric tensor g, the metric provides the notion of arc length, the differential arc length is given by:

$$ds = \sqrt{g_{\alpha\beta} dx^{\alpha} dx^{\beta}}$$

and the geodesic equation is a second-order differential equation in the coordinates, the general solution is a family of geodesics:

$$\frac{d^2 x^{\mu}}{ds^2} = -\Gamma^{\mu}_{\alpha\beta} \frac{dx^{\alpha}}{ds} \frac{dx^{\beta}}{ds}$$

where $\Gamma^{\mu}_{\alpha\beta}$ is a Christoffel symbol of the second kind, which contains the metric (with respect to the coordinate system).

Given the mass-energy distribution provided by the stress–energy tensor $T^{\alpha\beta}$, the Einstein field equations are a set of non-linear second-order partial differential equations in the metric, and imply the curvature of space time is equivalent to a gravitational field. Mass falling in curved spacetime is equivalent to a mass falling in a gravitational field - because gravity is a fictitious force. The *relative acceleration* of one geodesic to another in curved spacetime is given by the *geodesic deviation equation*:

$$\frac{D^2 \xi^{\alpha}}{ds^2} = -R^{\alpha}_{\beta\gamma\delta} \frac{dx^{\alpha}}{ds} \xi^{\gamma} \frac{dx^{\delta}}{ds}$$

where $\xi^{\alpha} = x_2{}^{\alpha} - x_1{}^{\alpha}$ is the separation vector between two geodesics, D/ds (*not* just d/ds) is the covariant derivative, and $R^{\alpha}_{\beta\gamma\delta}$ is the Riemann curvature tensor, containing the Christoffel symbols. In other words, the geodesic deviation equation is the equation of motion for masses in curved spacetime, analogous to the Lorentz force equation for charges in an electromagnetic field.

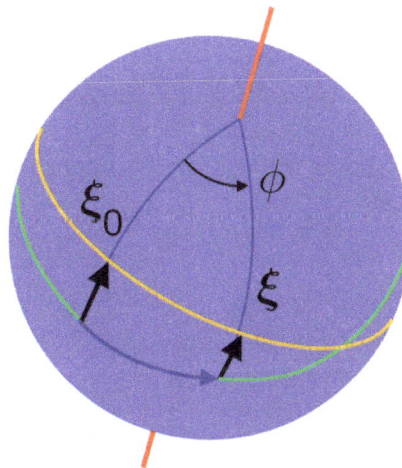

Geodesics on a sphere are arcs of great circles (yellow curve). On a 2D–manifold (such as the sphere shown), the direction of the accelerating geodesic is uniquely fixed if the separation vector ξ is orthogonal to the "fiducial geodesic" (green curve). As the separation vector ξ_0 changes to ξ after a distance s, the geodesics are not parallel (geodesic deviation).

For flat spacetime, the metric is a constant tensor so the Christoffel symbols vanish, and the geodesic equation has the solutions of straight lines. This is also the limiting case when masses move according to Newton's law of gravity.

Spinning Objects

In general relativity, rotational motion is described by the relativistic angular momentum tensor, including the spin tensor, which enter the equations of motion under covariant derivatives with respect to proper time. The Mathisson–Papapetrou–Dixon equations describe the motion of spinning objects moving in a gravitational field.

Analogues for Waves and Fields

Unlike the equations of motion for describing particle mechanics, which are systems of coupled ordinary differential equations, the analogous equations governing the dynamics of waves and fields are always partial differential equations, since the waves or fields are functions of space and time. For a particular solution, boundary conditions along with initial conditions need to be specified.

Sometimes in the following contexts, the wave or field equations are also called "equations of motion".

Field Equations

Equations that describe the spatial dependence and time evolution of fields are called *field equations*. These include

- Maxwell's equations for the electromagnetic field,

- Poisson's equation for Newtonian gravitational or electrostatic field potentials,

- the Einstein field equation for gravitation (Newton's law of gravity is a special case for weak gravitational fields and low velocities of particles).

This terminology is not universal: for example although the Navier–Stokes equations govern the velocity field of a fluid, they are not usually called "field equations", since in this context they represent the momentum of the fluid and are called the "momentum equations" instead.

Wave Equations

Equations of wave motion are called *wave equations*. The solutions to a wave equation give the time-evolution and spatial dependence of the amplitude. Boundary conditions determine if the solutions describe traveling waves or standing waves.

From classical equations of motion and field equations; mechanical, gravitational wave, and electromagnetic wave equations can be derived. The general linear wave equation in 3D is:

$$\frac{1}{v^2}\frac{\partial^2 X}{\partial t^2} = \nabla^2 X$$

where $X = X(r, t)$ is any mechanical or electromagnetic field amplitude, say:

- the transverse or longitudinal displacement of a vibrating rod, wire, cable, membrane etc.,

- the fluctuating pressure of a medium, sound pressure,

- the electric fields E or D, or the magnetic fields B or H,

- the voltage V or current I in an alternating current circuit,

and v is the phase velocity. Nonlinear equations model the dependence of phase velocity on amplitude, replacing v by $v(X)$. There are other linear and nonlinear wave equations for very specific applications.

Quantum Theory

In quantum theory, the wave and field concepts both appear.

In quantum mechanics, in which particles also have wave-like properties according to wave–particle duality, the analogue of the classical equations of motion (Newton's law, Euler–Lagrange equation, Hamilton–Jacobi equation, etc.) is the Schrödinger equation in its most general form:

$$i\hbar \frac{\partial \Psi}{\partial t} = \hat{H}\Psi,$$

where Ψ is the wavefunction of the system, $\hat{H}$ is the quantum Hamiltonian operator, rather than a function as in classical mechanics, and $\hbar$ is the Planck constant divided by 2π. Setting up the Hamiltonian and inserting it into the equation results in a wave equation, the solution is the wavefunction as a function of space and time. The Schrödinger equation itself reduces to the Hamilton–Jacobi equation when one considers the correspondence principle, in the limit that $\hbar$ becomes zero.

Throughout all aspects of quantum theory, relativistic or non-relativistic, there are various formulations alternative to the Schrödinger equation that govern the time evolution and behavior of a quantum system, for instance:

- the Heisenberg equation of motion resembles the time evolution of classical observables as functions of position, momentum, and time, if one replaces dynamical observables by their quantum operators and the classical Poisson bracket by the commutator,

- the phase space formulation closely follows classical Hamiltonian mechanics, placing position and momentum on equal footing,

- the Feynman path integral formulation extends the principle of least action to quantum mechanics and field theory, placing emphasis on the use of a Lagrangians rather than Hamiltonians.

References

- Torby, Bruce (1984). "Energy Methods". Advanced Dynamics for Engineers. HRW Series in Mechanical Engineering. United States of America: CBS College Publishing. ISBN 0-03-063366-4.

- Marion, JB; Thornton, ST (1995). Classical Dynamics of Systems and Particles (4th ed.). Thomson. ISBN 0-03-097302-3..

- Encyclopaedia of Physics (second Edition), R.G. Lerner, G.L. Trigg, VHC Publishers, 1991, ISBN (Verlags-gesellschaft) 3-527-26954-1 (VHC Inc.) 0-89573-752-3

- Analytical Mechanics, L.N. Hand, J.D. Finch, Cambridge University Press, 2008, ISBN 978-0-521-57572-0

- Halliday, David; Resnick, Robert; Walker, Jearl (2004-06-16). Fundamentals of Physics (7 Sub ed.). Wiley. ISBN 0-471-23231-9.

- M.R. Spiegel; S. Lipschutz; D. Spellman (2009). Vector Analysis. Schaum's Outlines (2nd ed.). McGraw Hill. p. 33. ISBN 978-0-07-161545-7.

- Essential Principles of Physics, P.M. Whelan, M.J. Hodgeson, second Edition, 1978, John Murray, ISBN 0-7195-3382-1

- Hanrahan, Val; Porkess, R (2003). Additional Mathematics for OCR. London: Hodder & Stoughton. p. 219. ISBN 0-340-86960-7.

- Keith Johnson (2001). Physics for you: revised national curriculum edition for GCSE (4th ed.). Nelson Thornes. p. 135. ISBN 978-0-7487-6236-1. The 5 symbols are remembered by "suvat". Given any three, the other two can be found.

- An Introduction to Mechanics, D. Kleppner, R.J. Kolenkow, Cambridge University Press, 2010, p. 112, ISBN 978-0-521-19821-9

- Encyclopaedia of Physics (second Edition), R.G. Lerner, G.L. Trigg, VHC publishers, 1991, ISBN (VHC Inc.) 0-89573-752-3

- Classical Mechanics (second Edition), T.W.B. Kibble, European Physics Series, Mc Graw Hill (UK), 1973, ISBN 0-07-084018-0.

- H.D. Young; R.A. Freedman (2008). University Physics (12th ed.). Addison-Wesley (Pearson International). ISBN 0-321-50130-6.

5

Sub-Disciplines of Mechanical Engineering

Mechatronics is a field of science that includes subjects such as computer engineering, telecommunications engineering, electronics and control engineering along with mechanical engineering. The various sub disciplines explained in this section are mechatronics, HVAC, robotics and tribology. This chapter will provide a glimpse of the sub disciplines of mechanical engineering.

Mechatronics

Mechatronics is a multidisciplinary field of science that includes a combination of mechanics, electronics, computer science, telecommunications science, systems science and control science. As technology advances, the subfields of engineering multiply and adapt. Mechatronics' aim is a design process that unifies these subfields. Originally, mechatronics just included the combination of mechanics and electronics, hence the word is a combination of mechanics and electronics; however, as technical systems have become more and more complex the definition has been broadened to include more technical areas.

The word "mechatronics" originated in Japanese-English and was created by Tetsuro Mori, an engineer of Yaskawa Electric Corporation. The word "mechatronics" was registered as trademark by the company in Japan with the registration number of "46-32714" in 1971. However, afterward the company released the right of using the word to public, and the word "mechatronics" spread to the rest of the world. Nowadays, the word is translated in each language and the word is considered as an essential term for industry.

French standard NF E 01-010 gives the following definition: "approach aiming at the synergistic integration of mechanics, electronics, control theory, and computer science within product design and manufacturing, in order to improve and/or optimize its functionality".

Many people treat "mechatronics" as a modern buzzword synonymous with "electromechanical engineering".

Description

A mechatronics engineer unites the principles of mechanics, electronics, and computing to generate a simpler, more economical and reliable system. The term "mechatronics" was coined by Tetsuro Mori, the senior engineer of the Japanese company Yaskawa in 1969. An industrial robot is a prime example of a mechatronics system; it includes aspects of electronics, mechanics, and computing to do its day-to-day jobs.

Engineering cybernetics deals with the question of control engineering of mechatronic systems. It is used to control or regulate such a system. Through collaboration, the mechatronic modules perform the production goals and inherit flexible and agile manufacturing properties in the production scheme. Modern production equipment consists of mechatronic modules that are integrated according to a control architecture. The most known architectures involve hierarchy, polyarchy, heterarchy, and hybrid. The methods for achieving a technical effect are described by control algorithms, which might or might not utilize formal methods in their design. Hybrid systems important to mechatronics include production systems, synergy drives, planetary exploration rovers, automotive subsystems such as anti-lock braking systems and spin-assist, and everyday equipment such as autofocus cameras, video, hard disks, and CD players.

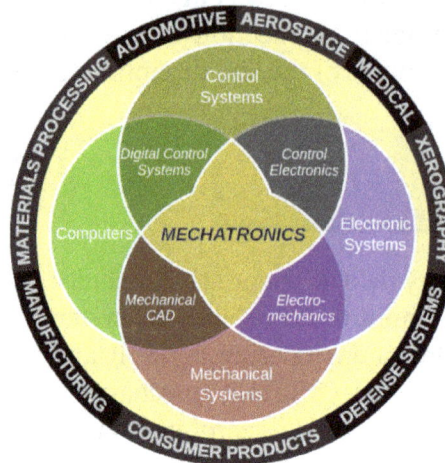

Aerial Euler diagram from RPI's website describes the s fields that make up Mechatronics

Course Structure

Mechatronic students take courses in various fields:

- Mechanical engineering and materials science
- Electrical engineering
- Computer engineering (software & hardware engineering)
- Computer science
- Systems and control engineering
- Optical engineering

Application

- Machine vision
- Automation and robotics
- Servo-mechanics
- Sensing and control systems
- Automotive engineering, automotive equipment in the design of subsystems such as anti-lock braking systems

- Computer-machine controls, such as computer driven machines like IE CNC milling machines
- Expert systems
- Industrial goods
- Consumer products
- Mechatronics systems
- Medical mechatronics, medical imaging systems
- Structural dynamic systems
- Transportation and vehicular systems
- Mechatronics as the new language of the automobile
- Computer aided and integrated manufacturing systems
- Computer-aided design
- Engineering and manufacturing systems
- Packaging
- Microcontrollers / PLCs
- Mobile apps
- M&E Engineering

Physical Implementations

Mechanical modeling calls for modeling and simulating physical complex phenomenon in the scope of a multi-scale and multi-physical approach. This implies to implement and to manage modeling and optimization methods and tools, which are integrated in a systemic approach. The specialty is aimed at students in mechanics who want to open their mind to systems engineering, and able to integrate different physics or technologies, as well as students in mechatronics who want to increase their knowledge in optimization and multidisciplinary simulation technics. The specialty educates students in robust and/or optimized conception methods for structures or many technological systems, and to the main modeling and simulation tools used in R&D. Special courses are also proposed for original applications (multi-materials composites, innovating transducers and actuators, integrated systems, ...) to prepare the students to the coming breakthrough in the domains covering the materials and the systems. For some mechatronic systems, the main issue is no longer how to implement a control system, but how to implement actuators. Within the mechatronic field, mainly two technologies are used to produce movement/motion.

Variant of The Field

An emerging variant of this field is biomechatronics, whose purpose is to integrate mechanical parts with a human being, usually in the form of removable gadgets such as an exoskeleton. This is the "real-life" version of cyberware.

Another variant that we can consider is Motion control for Advanced Mechatronics, which presently is recognized as a key technology in mechatronics. The robustness of motion control will be represented as a function of stiffness and a basis for practical realization. Target of motion is parameterized by control stiffness which could be variable according to the task reference. However, the system robustness of motion always requires very high stiffness in the controller.

HVAC

HVAC (heating, ventilating, and air conditioning; also heating, ventilation, and air conditioning) is the technology of indoor and vehicular environmental comfort. Its goal is to provide thermal comfort and acceptable indoor air quality. HVAC system design is a subdiscipline of mechanical engineering, based on the principles of thermodynamics, fluid mechanics, and heat transfer. Refrigeration is sometimes added to the field's abbreviation as HVAC&R or HVACR, or ventilating is dropped as in HACR (such as the designation of HACR-rated circuit breakers).

Rooftop HVAC unit with view of fresh air intake vent.

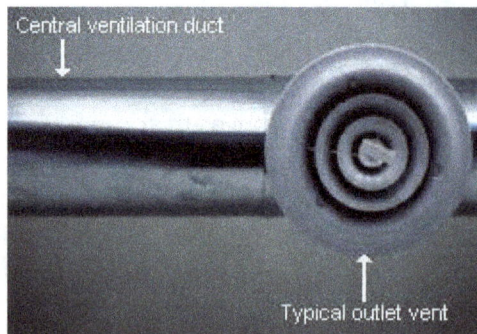

Ventilation duct with outlet vent. These are installed throughout a building to move air in or out of a room.

HVAC is important in the design of residential structures such as single family homes, apartment buildings, hotels and senior living facilities, medium to large industrial and office buildings such as skyscrapers and hospitals, onboard vessels, and in marine environments such as aquariums, where safe and healthy building conditions are regulated with respect to temperature and humidity, using fresh air from outdoors.

Ventilating or ventilation (the *V* in HVAC) is the process of exchanging or replacing air in any space to provide high indoor air quality which involves temperature control, oxygen replenishment, and removal of moisture, odors, smoke, heat, dust, airborne bacteria, carbon dioxide, and other gases. Ventilation removes unpleasant smells and excessive moisture, introduces outside air, keeps interior building air circulating, and prevents stagnation of the interior air.

Ventilation includes both the exchange of air to the outside as well as circulation of air within the building. It is one of the most important factors for maintaining acceptable indoor air quality in buildings. Methods for ventilating a building may be divided into *mechanical/forced* and *natural* types.

Overview

The three central functions of heating, ventilation, and air conditioning are interrelated, especially with the need to provide thermal comfort and acceptable indoor air quality within reasonable installation, operation, and maintenance costs. HVAC systems can provide ventilation, reduce air infiltration, and maintain pressure relationships between spaces. The means of air delivery and removal from spaces is known as room air distribution.

Individual Systems

In modern buildings the design, installation, and control systems of these functions are integrated into one or more HVAC systems. For very small buildings, contractors normally estimate the capacity, engineer, and select HVAC systems and equipment. For larger buildings, building service designers, mechanical engineers, or building services engineers analyze, design, and specify the HVAC systems. Specialty mechanical contractors then fabricate and commission the systems. Building permits and code-compliance inspections of the installations are normally required for all sizes of building.

District Networks

Although HVAC is executed in individual buildings or other enclosed spaces (like NORAD's underground headquarters), the equipment involved is in some cases an extension of a larger district heating (DH) or district cooling (DC) network, or a combined DHC network. In such cases, the operating and maintenance aspects are simplified and metering becomes necessary to bill for the energy that is consumed, and in some cases energy that is returned to the larger system. For example, at a given time one building may be utilizing chilled water for air conditioning and the warm water it returns may be used in another building for heating, or for the overall heating-portion of the DHC network (likely with energy added to boost the temperature).

Basing HVAC on a larger network helps to provide an economy of scale that is often not possible for individual buildings, for utilizing renewable energy sources such as solar heat, winter's cold, the cooling potential in some places of lakes or seawater for free cooling, and the enabling function of seasonal thermal energy storage.

History

HVAC is based on inventions and discoveries made by Nikolay Lvov, Michael Faraday, Willis Carrier, Edwin Ruud, Reuben Trane, James Joule, William Rankine, Sadi Carnot, and many others.

Multiple inventions within this time frame preceded the beginnings of first comfort air conditioning system, which was designed in 1902 by Alfred Wolff (Cooper, 2003) for the New York Stock Exchange, while Willis Carrier equipped the Sacketts-Wilhems Printing Company with the process AC unit the same year.

The invention of the components of HVAC systems went hand-in-hand with the industrial revolution, and new methods of modernization, higher efficiency, and system control are constantly being introduced by companies and inventors worldwide.

Heating

Heaters are appliances whose purpose is to generate heat (i.e. warmth) for the building. This can be done via central heating. Such a system contains a boiler, furnace, or heat pump to heat water, steam, or air in a central location such as a furnace room in a home, or a mechanical room in a large building. The heat can be transferred by convection, conduction, or radiation.

Generation

Central heating unit

Heaters exist for various types of fuel, including solid fuels, liquids, and gases. Another type of heat source is electricity, normally heating ribbons composed of high resistance wire. This principle is also used for baseboard heaters and portable heaters. Electrical heaters are often used as backup or supplemental heat for heat pump systems.

The heat pump gained popularity in the 1950s in Japan and the United States. Heat pumps can extract heat from various sources, such as environmental air, exhaust air from a building, or from the ground. Initially, heat pump HVAC systems were only used in moderate climates, but with improvements in low temperature operation and reduced loads due to more efficient homes, they are increasing in popularity in cooler climates.

Distribution

Water / Steam

In the case of heated water or steam, piping is used to transport the heat to the rooms. Most modern hot water boiler heating systems have a circulator, which is a pump, to move hot water through the distribution system (as opposed to older gravity-fed systems). The heat can be transferred to the surrounding air using radiators, hot water coils (hydro-air), or other heat exchangers. The radiators may be mounted on walls or installed within the floor to produce floor heat.

The use of water as the heat transfer medium is known as hydronics. The heated water can also supply an auxiliary heat exchanger to supply hot water for bathing and washing.

Air

Warm air systems distribute heated air through duct work systems of supply and return air through metal or fiberglass ducts. Many systems use the same ducts to distribute air cooled by an evaporator coil for air conditioning. The air supply is normally filtered through air cleaners to remove dust and pollen particles.

Dangers

The use of furnaces, space heaters, and boilers as a method of indoor heating could result in incomplete combustion and the emission of carbon monoxide, nitrogen oxides, formaldehyde, volatile organic compounds, and other combustion byproducts. Incomplete combustion occurs when there is insufficient oxygen; the inputs are fuels containing various contaminants and the outputs are harmful byproducts, most dangerously carbon monoxide, which is a tasteless and odorless gas with serious adverse health effects.

Without proper ventilation, carbon monoxide can be lethal at concentrations of 1000 ppm (0.1%). However, at several hundred ppm, carbon monoxide exposure induces headaches, fatigue, nausea, and vomiting. Carbon monoxide binds with hemoglobin in the blood, forming carboxyhemoglobin, reducing the blood's ability to transport oxygen. The primary health concerns associated with carbon monoxide exposure are its cardiovascular and neurobehavioral effects. Carbon monoxide can cause atherosclerosis (the hardening of arteries) and can also trigger heart attacks. Neurologically, carbon monoxide exposure reduces hand to eye coordination, vigilance, and continuous performance. It can also affect time discrimination.

Ventilation

Ventilation is the process of changing or replacing air in any space to control temperature or remove any combination of moisture, odors, smoke, heat, dust, airborne bacteria, or carbon dioxide, and to replenish oxygen. Ventilation includes both the exchange of air with the outside as well as circulation of air within the building. It is one of the most important factors for maintaining acceptable indoor air quality in buildings. Methods for ventilating a building may be divided into *mechanical/forced* and *natural* types.

Mechanical or Forced Ventilation

HVAC ventilation exhaust for a 12-story building

Mechanical, or forced, ventilation is provided by an air handler and used to control indoor air quality. Excess humidity, odors, and contaminants can often be controlled via dilution or replacement with outside air. However, in humid climates much energy is required to remove excess moisture from ventilation air.

Kitchens and bathrooms typically have mechanical exhausts to control odors and sometimes humidity. Factors in the design of such systems include the flow rate (which is a function of the fan speed and exhaust vent size) and noise level. Direct drive fans are available for many applications, and can reduce maintenance needs.

Ceiling fans and table/floor fans circulate air within a room for the purpose of reducing the perceived temperature by increasing evaporation of perspiration on the skin of the occupants. Because hot air rises, ceiling fans may be used to keep a room warmer in the winter by circulating the warm stratified air from the ceiling to the floor.

Natural Ventilation

Natural ventilation is the ventilation of a building with outside air without using fans or other mechanical systems. It can be via operable windows, louvers, or trickle vents when spaces are small and the architecture permits. In more complex schemes, warm air is allowed to rise and flow out high building openings to the outside (stack effect), causing cool outside air to be drawn into low building openings. Natural ventilation schemes can use very little energy, but care must be taken to ensure comfort. In warm or humid climates, maintaining thermal comfort solely via natural ventilation might not be possible. Air conditioning systems are used, either as backups or supplements. Air-side economizers also use outside air to condition spaces, but do so using fans, ducts, dampers, and control systems to introduce and distribute cool outdoor air when appropriate.

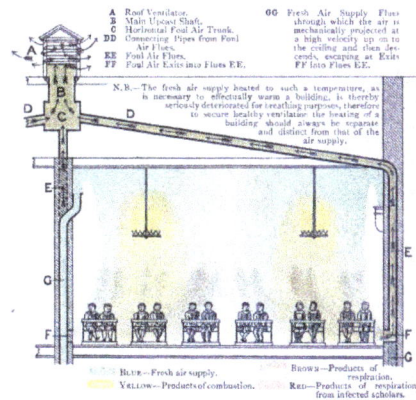

Ventilation on the downdraught system, by impulsion, or the 'plenum' principle, applied to schoolrooms (1899)

An important component of natural ventilation is air change rate or air changes per hour: the hourly rate of ventilation divided by the volume of the space. For example, six air changes per hour means an amount of new air, equal to the volume of the space, is added every ten minutes. For human comfort, a minimum of four air changes per hour is typical, though warehouses might have only two. Too high of an air change rate may be uncomfortable, akin to a wind tunnel which have thousands of changes per hour. The highest air change rates are for crowded spaces, bars, night clubs, commercial kitchens at around 30 to 50 air changes per hour.

Room pressure can be either positive or negative with respect to outside the room. Positive pressure occurs when there is more air being supplied than exhausted, and is common to reduce the infiltration of outside contaminants.

Airborne Diseases

Natural ventilation is a key factor in reducing the spread of airborne illnesses such as tuberculosis, the common cold, influenza and meningitis. Opening doors, windows, and using ceiling fans are all ways to maximize natural ventilation and reduce the risk of airborne contagion. Natural ventilation requires little maintenance and is inexpensive.

Air Conditioning

An air conditioning system, or a standalone air conditioner, provides cooling and humidity control for all or part of a building. Air conditioned buildings often have sealed windows, because open windows would work against the system intended to maintain constant indoor air conditions. Outside, fresh air is generally drawn into the system by a vent into the indoor heat exchanger section, creating positive air pressure. The percentage of return air made up of fresh air can usually be manipulated by adjusting the opening of this vent. Typical fresh air intake is about 10%.

Air conditioning and refrigeration are provided through the removal of heat. Heat can be removed through radiation, convection, or conduction. Refrigeration conduction media such as water, air, ice, and chemicals are referred to as refrigerants. A refrigerant is employed either in a heat pump system in which a compressor is used to drive thermodynamic refrigeration cycle, or in a free cooling system which uses pumps to circulate a cool refrigerant (typically water or a glycol mix).

Refrigeration Cycle

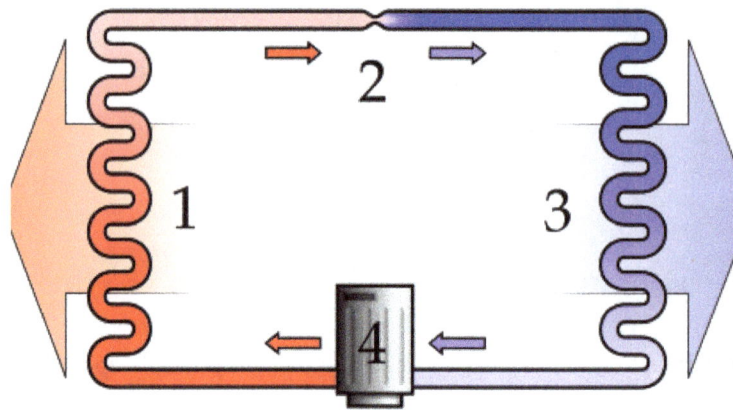

A simple stylized diagram of the refrigeration cycle: 1) condensing coil, 2) expansion valve, 3) evaporator coil, 4) compressor

The refrigeration cycle uses four essential elements to cool.

- The system refrigerant starts its cycle in a gaseous state. The compressor pumps the refrigerant gas up to a high pressure and temperature.

- From there it enters a heat exchanger (sometimes called a condensing coil or condenser) where it loses energy (heat) to the outside, cools, and condenses into its liquid phase.

- An expansion valve (also called metering device) regulates the refrigerant liquid to flow at the proper rate.

- The liquid refrigerant is returned to another heat exchanger where it is allowed to evaporate, hence the heat exchanger is often called an evaporating coil or evaporator. As the liquid refrigerant evaporates it absorbs energy (heat) from the inside air, returns to the compressor, and repeats the cycle. In the process, heat is absorbed from indoors and transferred outdoors, resulting in cooling of the building.

In variable climates, the system may include a reversing valve that switches from heating in winter to cooling in summer. By reversing the flow of refrigerant, the heat pump refrigeration cycle is changed from cooling to heating or vice versa. This allows a facility to be heated and cooled by a single piece of equipment by the same means, and with the same hardware.

Free Cooling

Free cooling systems can have very high efficiencies, and are sometimes combined with seasonal thermal energy storage so the cold of winter can be used for summer air conditioning. Common storage mediums are deep aquifers or a natural underground rock mass accessed via a cluster of small-diameter, heat-exchanger-equipped boreholes. Some systems with small storages are hybrids, using free cooling early in the cooling season, and later employing a heat pump to chill the circulation coming from the storage. The heat pump is added-in because the storage acts as a heat sink when the system is in cooling (as opposed to charging) mode, causing the temperature to gradually increase during the cooling season.

Some systems include an "economizer mode", which is sometimes called a "free-cooling mode". When economizing, the control system will open (fully or partially) the outside air damper and close (fully or partially) the return air damper. This will cause fresh, outside air to be supplied to the system. When the outside air is cooler than the demanded cool air, this will allow the demand to be met without using the mechanical supply of cooling (typically chilled water or a direct expansion "DX" unit), thus saving energy. The control system can compare the temperature of the outside air vs. return air, or it can compare the enthalpy of the air, as is frequently done in climates where humidity is more of an issue. In both cases, the outside air must be less energetic than the return air for the system to enter the economizer mode.

Central Vs. Split System

Central, "all-air" air-conditioning systems (or package systems) with a combined outdoor condenser/evaporator unit are often installed in modern residences, offices, and public buildings, but are difficult to retrofit (install in a building that was not designed to receive it) because of the bulky air ducts required. (Minisplit ductless systems are used in these situations.)

An alternative to central systems is the use of separate indoor and outdoor coils in split systems. These systems, although most often seen in residential applications, are gaining popularity in small commercial buildings. The evaporator coil is connected to a remote condenser unit using refrigerant piping between an indoor and outdoor unit instead of ducting air directly from the outdoor unit. Indoor units with directional vents mount onto walls, suspended from ceilings, or fit into the ceiling. Other indoor units mount inside the ceiling cavity, so that short lengths of duct handle air from the indoor unit to vents or diffusers around the rooms.

Dehumidification

Dehumidification (air drying) in an air conditioning system is provided by the evaporator. Since the evaporator operates at a temperature below the dew point, moisture in the air condenses on the evaporator coil tubes. This moisture is collected at the bottom of the evaporator in a pan and removed by piping to a central drain or onto the ground outside.

A dehumidifier is an air-conditioner-like device that controls the humidity of a room or building. It is often employed in basements which have a higher relative humidity because of their lower temperature (and propensity for damp floors and walls). In food retailing establishments, large open chiller cabinets are highly effective at dehumidifying the internal air. Conversely, a humidifier increases the humidity of a building.

Maintenance

All modern air conditioning systems, even small window package units, are equipped with internal air filters. These are generally of a lightweight gauzy material, and must be replaced or washed as conditions warrant. For example, a building in a high dust environment, or a home with furry pets, will need to have the filters changed more often than buildings without these dirt loads. Failure to replace these filters as needed will contribute to a lower heat exchange rate, resulting in wasted energy, shortened equipment life, and higher energy bills; low air flow can result in iced-over evaporator coils, which can completely stop air flow. Additionally, very dirty or plugged filters can cause overheating during a heating cycle, and can result in damage to the system or even fire.

Because an air conditioner moves heat between the indoor coil and the outdoor coil, both must be kept clean. This means that, in addition to replacing the air filter at the evaporator coil, it is also necessary to regularly clean the condenser coil. Failure to keep the condenser clean will eventually result in harm to the compressor, because the condenser coil is responsible for discharging both the indoor heat (as picked up by the evaporator) and the heat generated by the electric motor driving the compressor.

Energy Efficiency

Since the 1980s, manufacturers of HVAC equipment have been making an effort to make the systems they manufacture more efficient. This was originally driven by rising energy costs, and has more recently been driven by increased awareness of environmental issues. Additionally, improvements to the HVAC system efficiency can also help increase occupant health and productivity. In the US, the EPA has imposed tighter restrictions over the years. There are several methods for making HVAC systems more efficient.

Heating Energy

In the past, water heating was more efficient for heating buildings and was the standard in the United States. Today, forced air systems can double for air conditioning and are more popular.

Some benefits of forced air systems, which are now widely used in churches, schools and high-end residences, are

- Better air conditioning effects

- Energy savings of up to 15-20%

- Even conditioning

A drawback is the installation cost, which can be slightly higher than traditional HVAC systems.

Energy efficiency can be improved even more in central heating systems by introducing zoned heating. This allows a more granular application of heat, similar to non-central heating systems. Zones are controlled by multiple thermostats. In water heating systems the thermostats control zone valves, and in forced air systems they control zone dampers inside the vents which selectively block the flow of air. In this case, the control system is very critical to maintaining a proper temperature.

Forecasting is another method of controlling building heating by calculating demand for heating energy that should be supplied to the building in each time unit.

Geothermal Heat Pump

Geothermal heat pumps are similar to ordinary heat pumps, but instead of transferring heat to or from outside air, they rely on the stable, even temperature of the earth to provide heating and air conditioning. Many parts of the country experience seasonal temperature extremes, which would require large-capacity heating and cooling equipment to heat or cool buildings. For example, a conventional heat pump system used to heat a building in Montana's −70 °F (−57 °C) low temperature or cool a building in the highest temperature ever recorded in the US—134 °F (57 °C) in Death Valley, California, in 1913 would require a large amount of energy due to the extreme difference between inside and outside air temperatures. A few feet below the earth's surface, however, the ground remains at a relatively constant temperature. Utilizing this large source of relatively moderate temperature earth, a heating or cooling system's capacity can often be significantly reduced. Although ground temperatures vary according to latitude, at 6 feet (1.8 m) underground, temperatures generally only range from 45 to 75 °F (7 to 24 °C).

An example of a geothermal heat pump that uses a body of water as the heat sink, is the system used by the Trump International Hotel and Tower in Chicago, Illinois. This building is situated on the Chicago River, and uses cold river water by pumping it into a recirculating cooling system, where heat exchangers transfer heat from the building into the water, and then the now-warmed water is pumped back into the Chicago River.

While they may be more costly to install than regular heat pumps, geothermal heat pumps can produce markedly lower energy bills – 30 to 40 percent lower, according to estimates from the US Environmental Protection Agency.

Ventilation Energy Recovery

Energy recovery systems sometimes utilize heat recovery ventilation or energy recovery ventilation systems that employ heat exchangers or enthalpy wheels to recover sensible or latent heat from exhausted air. This is done by transfer of energy to the incoming outside fresh air.

Air Conditioning Energy

The performance of vapor compression refrigeration cycles is limited by thermodynamics. These air conditioning and heat pump devices *move* heat rather than convert it from one form to another, so *thermal efficiencies* do not appropriately describe the performance of these devices. The Coefficient-of-Performance (COP) measures performance, but this dimensionless measure has not been adopted. Instead, the Energy Efficiency Ratio (*EER*) has traditionally been used to characterize the performance of many HVAC systems. EER is the Energy Efficiency Ratio based on a 35 °C (95 °F) outdoor temperature. To more accurately describe the performance of air conditioning equipment over a typical cooling season a modified version of the EER, the Seasonal Energy Efficiency Ratio (*SEER*), or in Europe the ESEER, is used. SEER ratings are based on seasonal temperature averages instead of a constant 35 °C (95 °F) outdoor temperature. The current industry minimum SEER rating is 14 SEER.

Engineers have pointed out some areas where efficiency of the existing hardware could be improved. For example, the fan blades used to move the air are usually stamped from sheet metal, an economical method of manufacture, but as a result they are not aerodynamically efficient. A well-designed blade could reduce electrical power required to move the air by a third.

Air Filtration and Cleaning

Air cleaning and filtration removes particles, contaminants, vapors and gases from the air. The filtered and cleaned air then is used in heating, ventilation and air conditioning. Air cleaning and filtration should be taken in account when protecting our building environments.

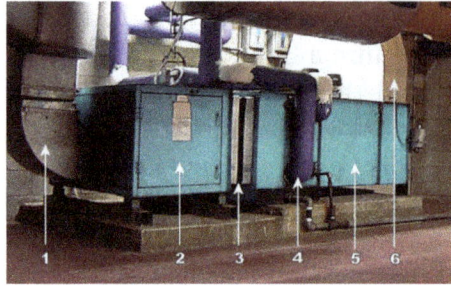
Air handling unit, used for heating, cooling, and filtering the air

Clean Air Delivery Rate and Filter Performance

Clean air delivery rate is the amount of clean air an air cleaner provides to a room or space. When determining CADR, the amount of airflow in a space is taken into account. For example, an air cleaner with a flow rate of 100 cfm (cubic feet per minute) and an efficiency of 50% has a CADR of 50 cfm. Along with CADR, filtration performance is very important when it comes to the air in our indoor environment. Filter performance depends on the size of the particle or fiber, the filter packing density and depth and also the air flow rate.

HVAC Industry and Standards

The HVAC industry is a worldwide enterprise, with roles including operation and maintenance, system design and construction, equipment manufacturing and sales, and in education and research. The HVAC industry was historically regulated by the manufacturers of HVAC equipment, but regulating and standards organizations such as HARDI, ASHRAE, SMACNA, ACCA, Uniform Mechanical Code, International Mechanical Code, and AMCA have been established to support the industry and encourage high standards and achievement.

The starting point in carrying out an estimate both for cooling and heating depends on the exterior climate and interior specified conditions. However, before taking up the heat load calculation, it is necessary to find fresh air requirements for each area in detail, as pressurization is an important consideration.

International

ISO 16813:2006 is one of the ISO building environment standards. It establishes the general principles of building environment design. It takes into account the need to provide a healthy indoor environment for the occupants as well as the need to protect the environment for future generations and promote collaboration among the various parties involved in building environmental design for sustainability. ISO16813 is applicable to new construction and the retrofit of existing buildings.

The building environmental design standard aims to:

- provide the constraints concerning sustainability issues from the initial stage of the design process, with building and plant life cycle to be considered together with owning and operating costs from the beginning of the design process;

- assess the proposed design with rational criteria for indoor air quality, thermal comfort, acoustical comfort, visual comfort, energy efficiency and HVAC system controls at every stage of the design process;

- iterate decisions and evaluations of the design throughout the design process.

North America

United States

In the United States, HVAC engineers generally are members of the American Society of Heating, Refrigerating, and Air-Conditioning Engineers (ASHRAE), EPA Universal CFC certified, or locally engineer certified such as a Special to Chief Boilers License issued by the state or, in some jurisdictions, the city. ASHRAE is an international technical society for all individuals and organizations interested in HVAC. The Society, organized into regions, chapters, and student branches, allows exchange of HVAC knowledge and experiences for the benefit of the field's practitioners and the public. ASHRAE provides many opportunities to participate in the development of new knowledge via, for example, research and its many technical committees. These committees typically meet twice per year at the ASHRAE Annual and Winter Meetings. A popular product show, the AHR Expo, is held in conjunction with each winter meeting. The Society has approximately 50,000 members and has headquarters in Atlanta, Georgia.

The most recognized standards for HVAC design are based on ASHRAE data. The most general of four volumes of the ASHRAE Handbook is Fundamentals; it includes heating and cooling calculations. Each volume of the ASHRAE Handbook is updated every four years. The design professional must consult ASHRAE data for the standards of design and care as the typical building codes provide little to no information on HVAC design practices; codes such as the UMC and IMC do include much detail on installation requirements, however.

American design standards are legislated in the Uniform Mechanical Code or International Mechanical Code. In certain states, counties, or cities, either of these codes may be adopted and amended via various legislative processes. These codes are updated and published by the International Association of Plumbing and Mechanical Officials (IAPMO) or the International Code Council (ICC) respectively, on a 3-year code development cycle. Typically, local building permit departments are charged with enforcement of these standards on private and certain public properties.

HVAC professionals in the US can receive training through formal training institutions, where most earn associate degrees. Training for HVAC technicians includes classroom lectures and hands-on tasks, and can be followed by an apprenticeship wherein the recent graduate works alongside a professional HVAC technician for a temporary period. HVAC techs who have been trained can also be certified in areas such as air conditioning, heat pumps, gas heating, and commercial refrigeration.

Europe

United Kingdom

The Chartered Institution of Building Services Engineers is a body that covers the essential Service (systems architecture) that allow buildings to operate. It includes the electrotechnical, heating, ventilating, air conditioning, refrigeration and plumbing industries. To train as a building services engineer, the academic requirements are GCSEs (A-C) / Standard Grades (1-3) in Maths and Science, which are important in measurements, planning and theory. Employers will often want a degree in a branch of engineering, such as building environment engineering, electrical engineering or mechanical engineering. To become a full member of CIBSE, and so also to be registered by the Engineering Council UK as a chartered engineer, engineers must also attain an Honours Degree and a master's degree in a relevant engineering subject.

CIBSE publishes several guides to HVAC design relevant to the UK market, and also the Republic of Ireland, Australia, New Zealand and Hong Kong. These guides include various recommended design criteria and standards, some of which are cited within the UK building regulations, and therefore form a legislative requirement for major building services works. The main guides are:

- Guide A: Environmental Design

- Guide B: Heating, Ventilating, Air Conditioning and Refrigeration

- Guide C: Reference Data

- Guide D: Transportation systems in Buildings

- Guide E: Fire Safety Engineering

- Guide F: Energy Efficiency in Buildings

- Guide G: Public Health Engineering

- Guide H: Building Control Systems

- Guide J: Weather, Solar and Illuminance Data

- Guide K: Electricity in Buildings

- Guide L: Sustainability

- Guide M: Maintenance Engineering and Management

Within the construction sector, it is the job of the building services engineer to design and oversee the installation and maintenance of the essential services such as gas, electricity, water, heating and lighting, as well as many others. These all help to make buildings comfortable and healthy places to live and work in. Building Services is part of a sector that has over 51,000 businesses and employs represents 2%-3% of the GDP.

Australia

The Air Conditioning and Mechanical Contractors Association of Australia (AMCA), Australian Institute of Refrigeration, Air Conditioning and Heating (AIRAH), and CIBSE are responsible.

Asia

Asian architectural temperature-control have different priorities than European methods. For example, Asian heating traditionally focuses on maintaining temperatures of objects such as the floor or furnishings such as Kotatsu tables and directly warming people, as opposed to the Western focus, in modern periods, on designing air systems.

Philippines

The Philippine Society of Ventilating, Air Conditioning and Refrigerating Engineers (PSVARE) along with Philippine Society of Mechanical Engineers (PSME) govern on the codes and standards for HVAC / MVAC in the Philippines.

India

The Indian Society of Heating, Refrigerating and Air Conditioning Engineers (ISHRAE) was established to promote the HVAC industry in India. ISHRAE is an associate of ASHRAE. ISHRAE was started at Delhi in 1981 and a chapter was started in Bangalore in 1989. Between 1989 & 1993, ISHRAE chapters were formed in all major cities in India and also in the Middle East.

Robotics

Robotics is the branch of mechanical engineering, electrical engineering and computer science that deals with the design, construction, operation, and application of robots, as well as computer systems for their control, sensory feedback, and information processing.

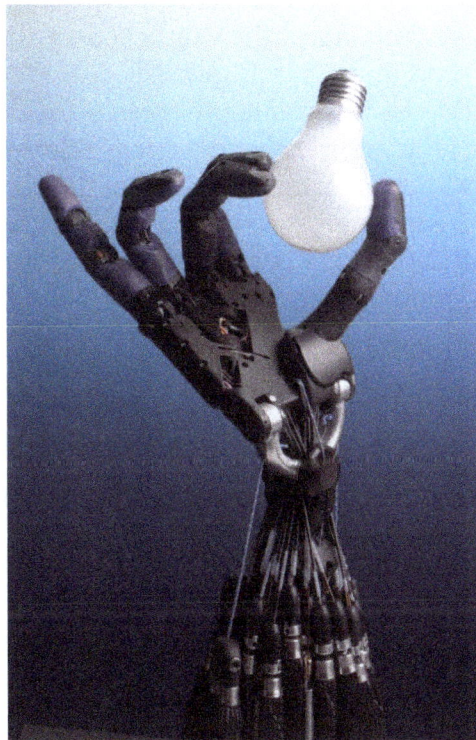

The Shadow robot hand system

These technologies deal with automated machines that can take the place of humans in dangerous environments or manufacturing processes, or resemble humans in appearance, behaviour, and or cognition. Many of today's robots are inspired by nature, contributing to the field of bio-inspired robotics.

The concept of creating machines that can operate autonomously dates back to classical times, but research into the functionality and potential uses of robots did not grow substantially until the 20th century. Throughout history, it has been frequently assumed that robots will one day be able to mimic human behavior and manage tasks in a human-like fashion. Today, robotics is a rapidly growing field, as technological advances continue; researching, designing, and building new robots serve various practical purposes, whether domestically, commercially, or militarily. Many robots are built to do jobs that are hazardous to people such as defusing bombs, finding survivors in unstable ruins, and exploring mines and shipwrecks. Robotics is also used in STEM (Science, Technology, Engineering, and Mathematics) as a teaching aid.

Etymology

The word *robotics* was derived from the word *robot*, which was introduced to the public by Czech writer Karel Čapek in his play *R.U.R. (Rossum's Universal Robots)*, which was published in 1920. The word *robot* comes from the Slavic word *robota*, which means labour. The play begins in a factory that makes artificial people called *robots*, creatures who can be mistaken for humans – very similar to the modern ideas of androids. Karel Čapek himself did not coin the word. He wrote a short letter in reference to an etymology in the *Oxford English Dictionary* in which he named his brother Josef Čapek as its actual originator.

According to the *Oxford English Dictionary*, the word *robotics* was first used in print by Isaac Asimov, in his science fiction short story "Liar!", published in May 1941 in *Astounding Science Fiction*. Asimov was unaware that he was coining the term; since the science and technology of electrical devices is *electronics*, he assumed *robotics* already referred to the science and technology of robots. In some of Asimov's other works, he states that the first use of the word *robotics* was in his short story *Runaround* (Astounding Science Fiction, March 1942). However, the original publication of "Liar!" predates that of "Runaround" by ten months, so the former is generally cited as the word's origin.

History of Robotics

In 1942 the science fiction writer Isaac Asimov created his Three Laws of Robotics.

In 1948 Norbert Wiener formulated the principles of cybernetics, the basis of practical robotics.

Fully autonomous robots only appeared in the second half of the 20th century. The first digitally operated and programmable robot, the Unimate, was installed in 1961 to lift hot pieces of metal from a die casting machine and stack them. Commercial and industrial robots are widespread today and used to perform jobs more cheaply, more accurately and more reliably, than humans. They are also employed in some jobs which are too dirty, dangerous, or dull to be suitable for humans. Robots are widely used in manufacturing, assembly, packing and packaging, transport, earth and space exploration, surgery, weaponry, laboratory research, safety, and the mass production of consumer and industrial goods.

Date	Significance	Robot Name	Inventor
Third century B.C. and earlier	One of the earliest descriptions of automata appears in the Lie Zi text, on a much earlier encounter between King Mu of Zhou (1023–957 BC) and a mechanical engineer known as Yan Shi, an 'artificer'. The latter allegedly presented the king with a life-size, human-shaped figure of his mechanical handiwork.		Yan Shi
First century A.D. and earlier	Descriptions of more than 100 machines and automata, including a fire engine, a wind organ, a coin-operated machine, and a steam-powered engine, in Pneumatica and Automata by Heron of Alexandria		Ctesibius, Philo of Byzantium, Heron of Alexandria, and others
c. 420 B.C.E	A wooden, steam propelled bird, which was able to fly		Archytas of Tarentum
1206	Created early humanoid automata, programmable automaton band	Robot band, hand-washing automaton, automated moving peacocks	Al-Jazari
1495	Designs for a humanoid robot	Mechanical Knight	Leonardo da Vinci
1738	Mechanical duck that was able to eat, flap its wings, and excrete	Digesting Duck	Jacques de Vaucanson
1898	Nikola Tesla demonstrates first radio-controlled vessel.	Teleautomaton	Nikola Tesla
1921	First fictional automatons called "robots" appear in the play R.U.R.	Rossum's Universal Robots	Karel Čapek
1930s	Humanoid robot exhibited at the 1939 and 1940 World's Fairs	Elektro	Westinghouse Electric Corporation
1946	First general-purpose digital computer	Whirlwind	Multiple people
1948	Simple robots exhibiting biological behaviors	Elsie and Elmer	William Grey Walter
1956	First commercial robot, from the Unimation company founded by George Devol and Joseph Engelberger, based on Devol's patents	Unimate	George Devol
1961	First installed industrial robot.	Unimate	George Devol
1973	First industrial robot with six electromechanically driven axes	Famulus	KUKA Robot Group
1974	The world's first microcomputer controlled electric industrial robot, IRB 6 from ASEA, was delivered to a small mechanical engineering company in southern Sweden. The design of this robot had been patented already 1972.	IRB 6	ABB Robot Group
1975	Programmable universal manipulation arm, a Unimation product	PUMA	Victor Scheinman

Robotic Aspects

Robotic construction

Electrical aspect

```
// Dev-C++ 4.9.9.2
// Project Type: Win32 GUI
// Window: Window Header
#include <Windows.h>
#include "resource.h"
// Window: Window Name
#ifdef NULL
#undef NULL
#define NULL 0
#endif
#define Wnd_Class "WIN_CHK"
#define Wnd_Title "預設碟密"
// Window: Window Parameters
static UINT WndPos_X = 0, WndPos_Y = 0;
// 400 x 300
static UINT WndPos_Width = 400, WndPos_Height = 300;
static HWND hwndWnd = 0;
static HINSTANCE hinstWnd = 0;
LRESULT CALLBACK WndProc(HWND, UINT, WPARAM, LPARAM);
BOOL ProcMsg(void);
BOOL BuildWnd( const char*, const char*);
void InitWindow_PositionCenter( UINT&, UINT&, UINT, UINT, BOOL);
// Window: Window Entry
int WINAPI WinMain(HINSTANCE hInstance, HINSTANCE hPrevInstance,
                   LPSTR lpCmdLine, int nShowCmd)
{
    //==START of WinMain==//
    if ( (hwndWnd = ::FindWindow( Wnd_Class, Wnd_Title)) != NULL )
    {
        ::SetForegroundWindow(hwndWnd);
        return NULL;
    }
    if ( BuildWnd( Wnd_Class, Wnd_Title) == TRUE )
    {
        while ( ProcMsg() == TRUE );
    }
    //==END of WinMain==//
    return NULL;
}
```

A level of programming

There are many types of robots; they are used in many different environments and for many different uses, although being very diverse in application and form they all share three basic similarities when it comes to their construction:

1. Robots all have some kind of mechanical construction, a frame, form or shape designed to achieve a particular task. For example, a robot designed to travel across heavy dirt or mud, might use caterpillar tracks. The mechanical aspect is mostly the creator's solution to completing the assigned task and dealing with the physics of the environment around it. Form follows function.

2. Robots have electrical components which power and control the machinery. For example, the robot with caterpillar tracks would need some kind of power to move the tracker treads. That power comes in the form of electricity, which will have to travel through a wire and originate from a battery, a basic electrical circuit. Even petrol powered machines that get their power mainly from petrol still require an electric current to start the combustion process which is why most petrol powered machines like cars, have batteries. The electrical aspect of robots is used for movement (through motors), sensing (where electrical signals are used to measure things like heat, sound, position, and energy status) and operation (robots need some level of electrical energy supplied to their motors and sensors in order to activate and perform basic operations)

3. All robots contain some level of computer programming code. A program is how a robot decides when or how to do something. In the caterpillar track example, a robot that needs to move across a muddy road may have the correct mechanical construction, and receive the correct amount of power from its battery, but would not go anywhere without a program telling it to move. Programs are the core essence of a robot, it could have excellent mechanical and electrical construction, but if its program is poorly constructed its performance will be very poor (or it may not perform at all). There are three different types of robotic programs: remote control, artificial intelligence and hybrid. A robot with remote control programing has a preexisting set of commands that it will only perform if and when it receives a signal from a control source, typically a human being with a remote control. It is perhaps more appropriate to view devices controlled primarily by human commands as falling in the discipline of automation rather than robotics. Robots that use artificial intelligence interact with their environment on their own without a control source, and can determine reactions to objects and problems they encounter using their preexisting programming. Hybrid is a form of programming that incorporates both AI and RC functions.

Applications

As more and more robots are designed for specific tasks this method of classification becomes more relevant. For example, many robots are designed for assembly work, which may not be readily adaptable for other applications. They are termed as "assembly robots". For seam welding, some suppliers provide complete welding systems with the robot i.e. the welding equipment along with other material handling facilities like turntables etc. as an integrated unit. Such an integrated robotic system is called a "welding robot" even though its discrete manipulator unit could be Va-dapted to a variety of tasks. Some robots are specifically designed for heavy load manipulation, and are labelled as "heavy duty robots."

Current and potential applications include:

- Military Robots

- Caterpillar plans to develop remote controlled machines and expects to develop fully autonomous heavy robots by 2021. Some cranes already are remote controlled.

- It was demonstrated that a robot can perform a herding task.

- Robots are increasingly used in manufacturing (since the 1960s). In the auto industry they can amount for more than half of the "labor". There are even "lights off" factories such as an IBM keyboard manufacturing factory in Texas that is 100% automated.

- Robots such as HOSPI are used as couriers in hospitals (hospital robot). Other hospital tasks performed by robots are receptionists, guides and porters helpers,

- Robots can serve as waiters and cooks., also at home. Boris is a robot that can load a dishwasher.

- Robot combat for sport – hobby or sport event where two or more robots fight in an arena to disable each other. This has developed from a hobby in the 1990s to several TV series worldwide.

- Cleanup of contaminated areas, such as toxic waste or nuclear facilities.

- Agricultural robots (AgRobots,).

- Domestic robots, cleaning and caring for the elderly

- Medical robots performing low-invasive surgery

- Household robots with full use.

- Nanorobots

Components

Power Source

At present mostly (lead–acid) batteries are used as a power source. Many different types of batteries can be used as a power source for robots. They range from lead–acid batteries, which are safe and have relatively long shelf lives but are rather heavy compared to silver–cadmium batteries that are much smaller in volume and are currently much more expensive. Designing a battery-powered robot needs to take into account factors such as safety, cycle lifetime and weight. Generators, often some type of internal combustion engine, can also be used. However, such designs are often mechanically complex and need fuel, require heat dissipation and are relatively heavy. A tether connecting the robot to a power supply would remove the power supply from the robot entirely. This has the advantage of saving weight and space by moving all power generation and storage components elsewhere. However, this design does come with the drawback of constantly having a cable connected to the robot, which can be difficult to manage. Potential power sources could be:

- pneumatic (compressed gases)

- Solar power (using the sun's energy and converting it into electrical power)

- hydraulics (liquids)

- flywheel energy storage

- organic garbage (through anaerobic digestion)

- nuclear

Actuation

Actuators are the "muscles" of a robot, the parts which convert stored energy into movement. By far the most popular actuators are electric motors that rotate a wheel or gear, and linear actuators that control industrial robots in factories. There are some recent advances in alternative types of actuators, powered by electricity, chemicals, or compressed air.

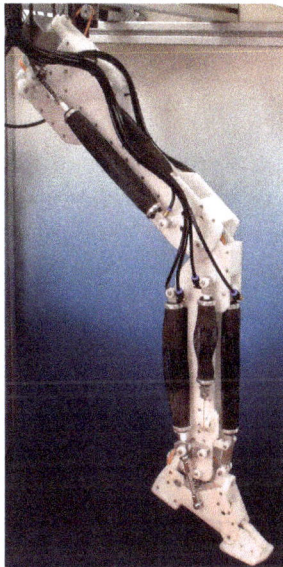

A robotic leg powered by air muscles

Electric Motors

The vast majority of robots use electric motors, often brushed and brushless DC motors in portable robots or AC motors in industrial robots and CNC machines. These motors are often preferred in systems with lighter loads, and where the predominant form of motion is rotational.

Linear Actuators

Various types of linear actuators move in and out instead of by spinning, and often have quicker direction changes, particularly when very large forces are needed such as with industrial robotics. They are typically powered by compressed air (pneumatic actuator) or an oil (hydraulic actuator).

Series Elastic Actuators

A spring can be designed as part of the motor actuator, to allow improved force control. It has been used in various robots, particularly walking humanoid robots.

Air Muscles

Pneumatic artificial muscles, also known as air muscles, are special tubes that expand(typically up to 40%) when air is forced inside them. They are used in some robot applications.

Muscle Wire

Muscle wire, also known as shape memory alloy, Nitinol® or Flexinol® wire, is a material which contracts (under 5%) when electricity is applied. They have been used for some small robot applications.

Electroactive Polymers

EAPs or EPAMs are a new plastic material that can contract substantially (up to 380% activation strain) from electricity, and have been used in facial muscles and arms of humanoid robots, and to enable new robots to float, fly, swim or walk.

Piezo Motors

Recent alternatives to DC motors are piezo motors or ultrasonic motors. These work on a fundamentally different principle, whereby tiny piezoceramic elements, vibrating many thousands of times per second, cause linear or rotary motion. There are different mechanisms of operation; one type uses the vibration of the piezo elements to step the motor in a circle or a straight line. Another type uses the piezo elements to cause a nut to vibrate or to drive a screw. The advantages of these motors are nanometer resolution, speed, and available force for their size. These motors are already available commercially, and being used on some robots.

Elastic Nanotubes

Elastic nanotubes are a promising artificial muscle technology in early-stage experimental development. The absence of defects in carbon nanotubes enables these filaments to deform elastically by several percent, with energy storage levels of perhaps 10 J/cm^3 for metal nanotubes. Human biceps could be replaced with an 8 mm diameter wire of this material. Such compact "muscle" might allow future robots to outrun and outjump humans.

Sensing

Sensors allow robots to receive information about a certain measurement of the environment, or internal components. This is essential for robots to perform their tasks, and act upon any changes in the environment to calculate the appropriate response. They are used for various forms of measurements, to give the robots warnings about safety or malfunctions, and to provide real time information of the task it is performing.

Touch

Current robotic and prosthetic hands receive far less tactile information than the human hand. Recent research has developed a tactile sensor array that mimics the mechanical properties and touch receptors of human fingertips. The sensor array is constructed as a rigid core surrounded by conductive fluid contained by an elastomeric skin. Electrodes are mounted on the surface of the rigid core and are connected to an impedance-measuring device within the core. When the artificial skin touches an object the fluid path around the electrodes is deformed, producing impedance changes that map the forces received from the object. The researchers expect that an important function of such artificial fingertips will be adjusting robotic grip on held objects.

Scientists from several European countries and Israel developed a prosthetic hand in 2009, called SmartHand, which functions like a real one—allowing patients to write with it, type on a keyboard, play piano and perform other fine movements. The prosthesis has sensors which enable the patient to sense real feeling in its fingertips.

Vision

Computer vision is the science and technology of machines that see. As a scientific discipline, computer vision is concerned with the theory behind artificial systems that extract information from images. The image data can take many forms, such as video sequences and views from cameras.

In most practical computer vision applications, the computers are pre-programmed to solve a particular task, but methods based on learning are now becoming increasingly common.

Computer vision systems rely on image sensors which detect electromagnetic radiation which is typically in the form of either visible light or infra-red light. The sensors are designed using solid-state physics. The process by which light propagates and reflects off surfaces is explained using optics. Sophisticated image sensors even require quantum mechanics to provide a complete understanding of the image formation process. Robots can also be equipped with multiple vision sensors to be better able to compute the sense of depth in the environment. Like human eyes, robots' "eyes" must also be able to focus on a particular area of interest, and also adjust to variations in light intensities.

There is a subfield within computer vision where artificial systems are designed to mimic the processing and behavior of biological system, at different levels of complexity. Also, some of the learning-based methods developed within computer vision have their background in biology.

Other

Other common forms of sensing in robotics use lidar, radar and sonar.

Manipulation

KUKA industrial robot operating in a foundry

Puma, one of the first industrial robots

Baxter, a modern and versatile industrial robot developed by Rodney Brooks

Robots need to manipulate objects; pick up, modify, destroy, or otherwise have an effect. Thus the "hands" of a robot are often referred to as *end effectors*, while the "arm" is referred to as a *manipulator*. Most robot arms have replaceable effectors, each allowing them to perform some small range of tasks. Some have a fixed manipulator which cannot be replaced, while a few have one very general purpose manipulator, for example a humanoid hand. Learning how to manipulate a robot often requires a close feedback between human to the robot, although there are several methods for remote manipulation of robots.

Mechanical Grippers

One of the most common effectors is the gripper. In its simplest manifestation it consists of just two fingers which can open and close to pick up and let go of a range of small objects. Fingers can for example be made of a chain with a metal wire run through it. Hands that resemble and work more like a human hand include the Shadow Hand and the Robonaut hand. Hands that are of a mid-level complexity include the Delft hand. Mechanical grippers can come in various types, including friction and encompassing jaws. Friction jaws use all the force of the gripper to hold the object in place using friction. Encompassing jaws cradle the object in place, using less friction.

Vacuum Grippers

Vacuum grippers are very simple astrictive devices, but can hold very large loads provided the prehension surface is smooth enough to ensure suction.

Pick and place robots for electronic components and for large objects like car windscreens, often use very simple vacuum grippers.

General Purpose Effectors

Some advanced robots are beginning to use fully humanoid hands, like the Shadow Hand, MANUS, and the Schunk hand. These are highly dexterous manipulators, with as many as 20 degrees of freedom and hundreds of tactile sensors.

Locomotion

Rolling Robots

For simplicity most mobile robots have four wheels or a number of continuous tracks. Some researchers have tried to create more complex wheeled robots with only one or two wheels. These can have certain advantages such as greater efficiency and reduced parts, as well as allowing a robot to navigate in confined places that a four-wheeled robot would not be able to.

Segway in the Robot museum in Nagoya

Two-wheeled Balancing Robots

Balancing robots generally use a gyroscope to detect how much a robot is falling and then drive the wheels proportionally in the same direction, to counterbalance the fall at hundreds of times per second, based on the dynamics of an inverted pendulum. Many different balancing robots have been designed. While the Segway is not commonly thought of as a robot, it can be thought of as a component of a robot, when used as such Segway refer to them as RMP (Robotic Mobility Platform). An example of this use has been as NASA's Robonaut that has been mounted on a Segway.

One-wheeled Balancing Robots

A one-wheeled balancing robot is an extension of a two-wheeled balancing robot so that it can move in any 2D direction using a round ball as its only wheel. Several one-wheeled balancing robots have been designed recently, such as Carnegie Mellon University's "Ballbot" that is the approximate height and width of a person, and Tohoku Gakuin University's "BallIP". Because of the long, thin shape and ability to maneuver in tight spaces, they have the potential to function better than other robots in environments with people.

Spherical Orb Robots

Several attempts have been made in robots that are completely inside a spherical ball, either by spinning a weight inside the ball, or by rotating the outer shells of the sphere. These have also been referred to as an orb bot or a ball bot.

Six-wheeled Robots

Using six wheels instead of four wheels can give better traction or grip in outdoor terrain such as on rocky dirt or grass.

Tracked Robots

TALON military robots used by the United States Army

Tank tracks provide even more traction than a six-wheeled robot. Tracked wheels behave as if they were made of hundreds of wheels, therefore are very common for outdoor and military robots, where the robot must drive on very rough terrain. However, they are difficult to use indoors such as on carpets and smooth floors. Examples include NASA's Urban Robot "Urbie".

Walking Applied to Robots

Walking is a difficult and dynamic problem to solve. Several robots have been made which can walk reliably on two legs, however none have yet been made which are as robust as a human. There has been much study on human inspired walking, such as AMBER lab which was established in 2008 by the Mechanical Engineering Department at Texas A&M University. Many other robots have been built that walk on more than two legs, due to these robots being significantly easier to construct. Walking robots can be used for uneven terrains, which would provide better mobility and energy efficiency than other locomotion methods. Hybrids too have been proposed in movies such as I, Robot, where they walk on 2 legs and switch to 4 (arms+legs) when going to a sprint. Typically, robots on 2 legs can walk well on flat floors and can occasionally walk up stairs. None can walk over rocky, uneven terrain. Some of the methods which have been tried are:

ZMP Technique

The Zero Moment Point (ZMP) is the algorithm used by robots such as Honda's ASIMO. The robot's onboard computer tries to keep the total inertial forces (the combination of Earth's gravity and the acceleration and deceleration of walking), exactly opposed by the floor reaction force (the force of the floor pushing back on the robot's foot). In this way, the two forces cancel out, leaving no moment (force causing the robot to rotate and fall over). However, this is not exactly how a human walks, and the difference is obvious to human observers, some of whom have pointed out that ASIMO walks as if it needs the lavatory. ASIMO's walking algorithm is not static, and some dynamic balancing is used. However, it still requires a smooth surface to walk on.

Hopping

Several robots, built in the 1980s by Marc Raibert at the MIT Leg Laboratory, successfully demonstrated very dynamic walking. Initially, a robot with only one leg, and a very small foot, could stay upright simply by hopping. The movement is the same as that of a person on a pogo stick. As the robot falls to one side, it would jump slightly in that direction, in order to catch itself. Soon, the algorithm was generalised to two and four legs. A bipedal robot was demonstrated running and even performing somersaults. A quadruped was also demonstrated which could trot, run, pace, and bound.

Dynamic Balancing (Controlled Falling)

A more advanced way for a robot to walk is by using a dynamic balancing algorithm, which is potentially more robust than the Zero Moment Point technique, as it constantly monitors the robot's motion, and places the feet in order to maintain stability. This technique was recently demonstrated by Anybots' Dexter Robot, which is so stable, it can even jump. Another example is the TU Delft Flame.

Passive Dynamics

Perhaps the most promising approach utilizes passive dynamics where the momentum of swinging limbs is used for greater efficiency. It has been shown that totally unpowered humanoid mechanisms can walk down a gentle slope, using only gravity to propel themselves. Using this technique, a robot need only supply a small amount of motor power to walk along a flat surface or a little more to walk up a hill. This technique promises to make walking robots at least ten times more efficient than ZMP walkers, like ASIMO.

Other Methods of Locomotion

Flying

A modern passenger airliner is essentially a flying robot, with two humans to manage it. The autopilot can control the plane for each stage of the journey, including takeoff, normal flight, and even landing. Other flying robots are uninhabited, and are known as unmanned aerial vehicles (UAVs). They can be smaller and lighter without a human pilot on board, and fly into dangerous territory for military surveillance missions. Some can even fire on targets under command. UAVs are also being developed which can fire on targets automatically, without the need for a command from a human. Other flying robots include cruise missiles, the Entomopter, and the Epson micro heli-copter robot. Robots such as the Air Penguin, Air Ray, and Air Jelly have lighter-than-air bodies, propelled by paddles, and guided by sonar.

Two robot snakes. Left one has 64 motors (with 2 degrees of freedom per segment), the right one 10.

Snaking

Several snake robots have been successfully developed. Mimicking the way real snakes move, these robots can navigate very confined spaces, meaning they may one day be used to search for people trapped in collapsed buildings. The Japanese ACM-R5 snake robot can even navigate both on land and in water.

Skating

A small number of skating robots have been developed, one of which is a multi-mode walking and skating device. It has four legs, with unpowered wheels, which can either step or roll. Another robot, Plen, can use a miniature skateboard or roller-skates, and skate across a desktop.

Capuchin, a climbing robot

Climbing

Several different approaches have been used to develop robots that have the ability to climb vertical surfaces. One approach mimics the movements of a human climber on a wall with protrusions; adjusting the center of mass and moving each limb in turn to gain leverage. An example of this is Capuchin, built by Dr. Ruixiang Zhang at Stanford University, California. Another approach uses the specialized toe pad method of wall-climbing geckoes, which can run on smooth surfaces such as vertical glass. Examples of this approach include Wallbot and Stickybot. China's *Technology Daily* reported on November 15, 2008 that Dr. Li Hiu Yeung and his research group of New Concept Aircraft (Zhuhai) Co., Ltd. had successfully developed a bionic gecko robot named "Speedy Freelander". According to Dr. Li, the gecko robot could rapidly climb up and down a variety of building walls, navigate through ground and wall fissures, and walk upside-down on the ceiling. It was also able to adapt to the surfaces of smooth glass, rough, sticky or dusty walls as well as various types of metallic materials. It could also identify and circumvent obstacles automatically. Its flexibility and speed were comparable to a natural gecko. A third approach is to mimic the motion of a snake climbing a pole.. Lastly one may mimic the movements of a human climber on a wall with protrusions; adjusting the center of mass and moving each limb in turn to gain leverage.

Swimming (Piscine)

It is calculated that when swimming some fish can achieve a propulsive efficiency greater than 90%. Furthermore, they can accelerate and maneuver far better than any man-made boat or submarine, and produce less noise and water disturbance. Therefore, many researchers studying underwater robots would like to copy this type of locomotion. Notable examples are the Essex University Computer Science Robotic Fish G9, and the Robot Tuna built by the Institute of Field Robotics, to analyze and mathematically model thunniform motion. The Aqua Penguin, designed and built by Festo of Germany, copies the streamlined shape and propulsion by front "flippers" of penguins. Festo have also built the Aqua Ray and Aqua Jelly, which emulate the locomotion of manta ray, and jellyfish, respectively.

Robotic Fish: *iSplash*-II

In 2014 *iSplash*-II was developed by R.J Clapham PhD at Essex University. It was the first robotic fish capable of outperforming real carangiform fish in terms of average maximum velocity (measured in body lengths/ second) and endurance, the duration that top speed is maintained. This build attained swimming speeds of 11.6BL/s (i.e. 3.7 m/s). The first build, *iSplash*-I (2014) was the first robotic platform to apply a full-body length carangiform swimming motion which was found to increase swimming speed by 27% over the traditional approach of a posterior confined wave form.

Sailing

The autonomous sailboat robot *Vaimos*

Sailboat robots have also been developed in order to make measurements at the surface of the ocean. A typical sailboat robot is *Vaimos* built by IFREMER and ENSTA-Bretagne. Since the propulsion of sailboat robots uses the wind, the energy of the batteries is only used for the computer, for the communication and for the actuators (to tune the rudder and the sail). If the robot is equipped with solar panels, the robot could theoretically navigate forever. The two main competitions of sailboat robots are WRSC, which takes place every year in Europe, and Sailbot.

Environmental Interaction and Navigation

Radar, GPS, and lidar, are all combined to provide proper navigation and obstacle avoidance (vehicle developed for 2007 DARPA Urban Challenge)

Though a significant percentage of robots in commission today are either human controlled, or operate in a static environment, there is an increasing interest in robots that can operate autonomously in a dynamic environment. These robots require some combination of navigation hardware and software in order to traverse their environment. In particular unforeseen events (e.g. people and other obstacles that are not stationary) can cause problems or collisions. Some highly advanced robots such as ASIMO, and Meinü robot have particularly good robot navigation hardware and software. Also, self-controlled cars, Ernst Dickmanns' driverless car, and the entries in the DARPA Grand Challenge, are capable of sensing the environment well and subsequently making navigational decisions based on this information. Most of these robots employ a GPS navigation device with waypoints, along with radar, sometimes combined with other sensory data such as lidar, video cameras, and inertial guidance systems for better navigation between waypoints.

Human-Robot Interaction

Kismet can produce a range of facial expressions.

The state of the art in sensory intelligence for robots will have to progress through several orders of magnitude if we want the robots working in our homes to go beyond vacuum-cleaning the floors. If robots are to work effectively in homes and other non-industrial environments, the way they are instructed to perform their jobs, and especially how they will be told to stop will be of critical importance. The people who interact with them may have little or no training in robotics, and so any interface will need to be extremely intuitive. Science fiction authors also typically assume that robots will eventually be capable of communicating with humans through speech, gestures, and facial expressions, rather than a command-line interface. Although speech would be the most natural way for the human to communicate, it is unnatural for the robot. It will probably be a long time before robots interact as naturally as the fictional C-3PO, or Data of Star Trek, Next Generation.

Speech Recognition

Interpreting the continuous flow of sounds coming from a human, in real time, is a difficult task for a computer, mostly because of the great variability of speech. The same word, spoken by the same person may sound different depending on local acoustics, volume, the previous word, whether or not the speaker has a cold, etc.. It becomes even harder when the speaker has a different accent. Nevertheless, great strides have been made in the field since Davis, Biddulph, and Balashek designed the first "voice input system" which recognized "ten digits spoken by a single user with 100% accuracy" in 1952. Currently, the best systems can recognize continuous, natural speech, up to 160 words per minute, with an accuracy of 95%.

Robotic Voice

Other hurdles exist when allowing the robot to use voice for interacting with humans. For social reasons, synthetic voice proves suboptimal as a communication medium, making it necessary to develop the emotional component of robotic voice through various techniques.

Gestures

One can imagine, in the future, explaining to a robot chef how to make a pastry, or asking directions from a robot police officer. In both of these cases, making hand gestures would aid the verbal descriptions. In the first case, the robot would be recognizing gestures made by the human, and perhaps repeating them for confirmation. In the second case, the robot police officer would gesture to indicate "down the road, then turn right". It is likely that gestures will make up a part of the interaction between humans and robots. A great many systems have been developed to recognize human hand gestures.

Facial Expression

Facial expressions can provide rapid feedback on the progress of a dialog between two humans, and soon may be able to do the same for humans and robots. Robotic faces have been constructed by Hanson Robotics using their elastic polymer called Frubber, allowing a large number of facial expressions due to the elasticity of the rubber facial coating and embedded subsurface motors (servos). The coating and servos are built on a metal skull. A robot should know how to approach a human, judging by their facial expression and body language. Whether the person is happy, frightened, or crazy-looking affects the type of interaction expected of the robot. Likewise, robots like Kismet and the more recent addition, Nexi can produce a range of facial expressions, allowing it to have meaningful social exchanges with humans.

Artificial Emotions

Artificial emotions can also be generated, composed of a sequence of facial expressions and/or gestures. As can be seen from the movie Final Fantasy: The Spirits Within, the programming of these artificial emotions is complex and requires a large amount of human observation. To simplify this programming in the movie, presets were created together with a special software program. This decreased the amount of time needed to make the film. These presets could possibly be transferred for use in real-life robots.

Personality

Many of the robots of science fiction have a personality, something which may or may not be desirable in the commercial robots of the future. Nevertheless, researchers are trying to create robots which appear to have a personality: i.e. they use sounds, facial expressions, and body language to try to convey an internal state, which may be joy, sadness, or fear. One commercial example is Pleo, a toy robot dinosaur, which can exhibit several apparent emotions.

Social Intelligence

The Socially Intelligent Machines Lab of the Georgia Institute of Technology researches new concepts of guided teaching interaction with robots. Aim of the projects is a social robot learns task goals from human demonstrations without prior knowledge of high-level concepts. These new concepts are grounded from low-level continuous sensor data through unsupervised learning, and task goals are subsequently learned using a Bayesian approach. These concepts can be used to transfer knowledge to future tasks, resulting in faster learning of those tasks. The results are demonstrated by the robot *Curi* who can scoop some pasta from a pot onto a plate and serve the sauce on top.

Control

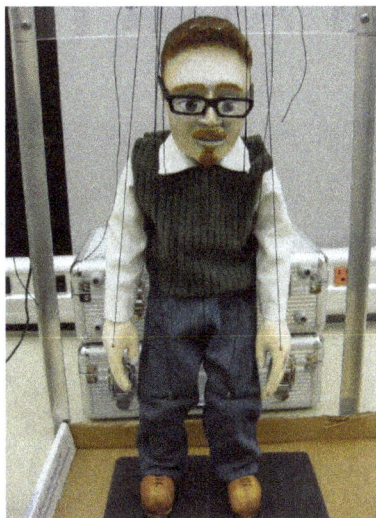

Puppet Magnus, a robot-manipulated marionette with complex control systems

RuBot II can resolve manually Rubik cubes

The mechanical structure of a robot must be controlled to perform tasks. The control of a robot involves three distinct phases – perception, processing, and action (robotic paradigms). Sensors give information about the environment or the robot itself (e.g. the position of its joints or its end effector). This information is then processed to be stored or transmitted, and to calculate the appropriate signals to the actuators (motors) which move the mechanical.

The processing phase can range in complexity. At a reactive level, it may translate raw sensor information directly into actuator commands. Sensor fusion may first be used to estimate parameters of interest (e.g. the position of the robot's gripper) from noisy sensor data. An immediate task (such as moving the gripper in a certain direction) is inferred from these estimates. Techniques from control theory convert the task into commands that drive the actuators.

At longer time scales or with more sophisticated tasks, the robot may need to build and reason with a "cognitive" model. Cognitive models try to represent the robot, the world, and how they interact. Pattern recognition and computer vision can be used to track objects. Mapping techniques can be used to build maps of the world. Finally, motion planning and other artificial intelligence techniques may be used to figure out how to act. For example, a planner may figure out how to achieve a task without hitting obstacles, falling over, etc.

Autonomy Levels

TOPIO, a humanoid robot, played ping pong at Tokyo IREX 2009.

Control systems may also have varying levels of autonomy.

1. Direct interaction is used for haptic or tele-operated devices, and the human has nearly complete control over the robot's motion.

2. Operator-assist modes have the operator commanding medium-to-high-level tasks, with the robot automatically figuring out how to achieve them.

3. An autonomous robot may go for extended periods of time without human interaction. Higher levels of autonomy do not necessarily require more complex cognitive capabilities. For example, robots in assembly plants are completely autonomous, but operate in a fixed pattern.

Another classification takes into account the interaction between human control and the machine motions.

1. Teleoperation. A human controls each movement, each machine actuator change is specified by the operator.

2. Supervisory. A human specifies general moves or position changes and the machine decides specific movements of its actuators.

3. Task-level autonomy. The operator specifies only the task and the robot manages itself to complete it.

4. Full autonomy. The machine will create and complete all its tasks without human interaction.

Robotics Research

Much of the research in robotics focuses not on specific industrial tasks, but on investigations into new types of robots, alternative ways to think about or design robots, and new ways to manufacture them but other investigations, such as MIT's cyberflora project, are almost wholly academic.

A first particular new innovation in robot design is the opensourcing of robot-projects. To describe the level of advancement of a robot, the term "Generation Robots" can be used. This term is coined by Professor Hans Moravec, Principal Research Scientist at the Carnegie Mellon University Robotics Institute in describing the near future evolution of robot technology. *First generation* robots, Moravec predicted in 1997, should have an intellectual capacity comparable to perhaps a lizard and should become available by 2010. Because the *first generation* robot would be incapable of learning, however, Moravec predicts that the *second generation* robot would be an improvement over the *first* and become available by 2020, with the intelligence maybe comparable to that of a mouse. The *third generation* robot should have the intelligence comparable to that of a monkey. Though *fourth generation* robots, robots with human intelligence, professor Moravec predicts, would become possible, he does not predict this happening before around 2040 or 2050.

The second is evolutionary robots. This is a methodology that uses evolutionary computation to help design robots, especially the body form, or motion and behavior controllers. In a similar way to natural evolution, a large population of robots is allowed to compete in some way, or their ability to perform a task is measured using a fitness function. Those that perform worst are removed from the population, and replaced by a new set, which have new behaviors based on those of the winners. Over time the population improves, and eventually a satisfactory robot may appear. This happens without any direct programming of the robots by the researchers. Researchers use this method both to create better robots, and to explore the nature of evolution. Because the process often requires many generations of robots to be simulated, this technique may be run entirely or mostly in simulation, then tested on real robots once the evolved algorithms are good enough. Currently, there are about 10 million industrial robots toiling around the world, and Japan is the top country having high density of utilizing robots in its manufacturing industry.

Dynamics and Kinematics

The study of motion can be divided into kinematics and dynamics. Direct kinematics refers to the calculation of end effector position, orientation, velocity, and acceleration when the corresponding joint values are known. Inverse kinematics refers to the opposite case in which required joint values are calculated for given end effector values, as done in path planning. Some special aspects of kinematics include handling of redundancy (different possibilities of performing the same movement), collision avoidance, and singularity avoidance. Once all relevant positions, velocities, and accelerations have been calculated using kinematics, methods from the field of dynamics are used to study the effect of forces upon these movements. Direct dynamics refers to the calculation of accelerations in the robot once the applied forces are known. Direct dynamics is used in computer simulations of the robot. Inverse dynamics refers to the calculation of the actuator forces necessary to create a prescribed end effector acceleration. This information can be used to improve the control algorithms of a robot.

In each area mentioned above, researchers strive to develop new concepts and strategies, improve existing ones, and improve the interaction between these areas. To do this, criteria for "optimal" performance and ways to optimize design, structure, and control of robots must be developed and implemented.

Bionics and Biomimetics

Bionics and biomimetics apply the physiology and methods of locomotion of animals to the design of robots. For example, the design of BionicKangaroo was based on the way kangaroos jump.

Education and Training

Robotics engineers design robots, maintain them, develop new applications for them, and conduct research to expand the potential of robotics. Robots have become a popular educational tool in some middle and high schools, particularly in parts of the USA, as well as in numerous youth summer camps, raising interest in programming, artificial intelligence and robotics among students. First-year computer science courses at some universities now include programming of a robot in addition to traditional software engineering-based coursework.

The SCORBOT-ER 4u educational robot

Career Training

Universities offer bachelors, masters, and doctoral degrees in the field of robotics. Vocational schools offer robotics training aimed at careers in robotics.

Certification

The Robotics Certification Standards Alliance (RCSA) is an international robotics certification authority that confers various industry- and educational-related robotics certifications.

Summer Robotics Camp

Several national summer camp programs include robotics as part of their core curriculum. In addition, youth summer robotics programs are frequently offered by celebrated museums such as the American Museum of Natural History and The Tech Museum of Innovation in Silicon Valley, CA, just to name a few.

Robotics Competitions

There are lots of competitions all around the globe. One of the most important competitions is the FLL or FIRST Lego League. The idea of this specific competition is that kids start developing knowledge and getting into robotics while playing with Legos since they are 9 years old. This competition is associated with Ni or National Instruments.

Robotics Afterschool Programs

Many schools across the country are beginning to add robotics programs to their after school curriculum. Some major programs for afterschool robotics include FIRST Robotics Competition, Botball and B.E.S.T. Robotics. Robotics competitions often include aspects of business and marketing as well as engineering and design.

The Lego company began a program for children to learn and get excited about robotics at a young age.

Employment

A robot technician builds small all-terrain robots. (Courtesy: MobileRobots Inc)

Robotics is an essential component in many modern manufacturing environments. As factories increase their use of robots, the number of robotics–related jobs grow and have been observed to be steadily rising. The employment of robots in industries has increased productivity and efficiency savings and is typically seen as a long term investment for benefactors.

Occupational Safety and Health Implications of Robotics

A discussion paper drawn up by EU-OSHA highlights how the spread of robotics presents both opportunities and challenges for occupational safety and health (OSH).

The greatest OSH benefits stemming from the wider use of robotics should be substitution for people working in unhealthy or dangerous environments. In space, defence, security, or the nuclear industry, but also in logistics, maintenance and inspection, autonomous robots are particularly useful in replacing human workers performing dirty, dull or unsafe tasks, thus avoiding workers' exposures to hazardous agents and conditions and reducing physical, ergonomic and psychosocial risks. For example, robots are already used to perform repetitive and monotonous tasks, to handle radioactive material or to work in explosive atmospheres. In the future, many other highly repetitive, risky or unpleasant tasks will be performed by robots in a variety of sectors like agriculture, construction, transport, healthcare, firefighting or cleaning services.

Despite these advances, there are certain skills to which humans will be better suited than machines for some time to come and the question is how to achieve the best combination of human and robot skills. The advantages of robotics include heavy-duty jobs with precision and repeatability, whereas the advantages of humans include creativity, decision-making, flexibility and adaptability. This need to combine optimal skills has resulted in collaborative robots and humans sharing a common workspace more closely and led to the development of new approaches and standards to guarantee the safety of the "man-robot merger". Some European countries are including robotics in their national programmes and trying to promote a safe and flexible co-operation between robots and operators to achieve better productivity. For example, the German Federal Institute for Occupational Safety and Health (BAuA) organises annual workshops on the topic "human-robot collaboration".

In future, co-operation between robots and humans will be diversified, with robots increasing their autonomy and human-robot collaboration reaching completely new forms. Current approaches and technical standards aiming to protect employees from the risk of working with collaborative robots will have to be revised.

Tribology

Tribology is the study of science and engineering of interacting surfaces in relative motion. It includes the study and application of the principles of friction, lubrication and wear. Tribology is a branch of mechanical engineering and materials science.

Etymology

It was coined by the British physicist David Tabor, and also by Peter Jost in 1964, a lubrication expert who noticed the problems with increasing friction on machines, and started the new discipline of tribology.

Fundamentals

The tribological interactions of a solid surface's exposed face with interfacing materials and environment may result in loss of material from the surface. The process leading to loss of material is known as "wear". Major types of wear include abrasion, friction (adhesion and cohesion), erosion, and corrosion. Wear can be minimized by modifying the surface properties of the solids by one or more "surface engineering" processes (also called surface finishing) or by use of lubricants (for frictional or adhesive wear).

Estimated direct and consequential annual loss to industries in the USA due to wear is approximately 1-2% of GDP. (Heinz, 1987). Engineered surfaces extend the working life of both original and recycled and resurfaced equipment, thus saving large sums of money and leading to conservation of material, energy and the environment. Methodologies to minimize wear include systematic approaches to diagnose the wear and to prescribe appropriate solutions. Important methods include:

- Point like contact theory was established by Heinrich Hertz in 1880s.

- Fluid lubrication dynamics was established by Arnold Johannes Sommerfeld in 1900s.

- Terotechnology, where multidisciplinary engineering and management techniques are used to protect equipment and machinery from degradation (Peter Jost, 1972)

- Horst Czichos's systems approach, where appropriate material is selected by checking properties against tribological requirements under operating environment (H. Czichos,1978)

- Asset Management by Material Prognosis - a concept similar to terotechnology which has been introduced by the US Military (DARPA) for upkeep of equipment in good health and start-ready condition for 24 hours. Good health monitoring systems combined with appropriate remedies at maintenance and repair stages have led to improved performance, reliability and extended life cycle of the assets, such as advanced military hardware and civil aircraft.

In recent years, micro- and nanotribology have been gaining ground. Frictional interactions in microscopically small components are becoming increasingly important for the development of new products in electronics, life sciences, chemistry, sensors and by extension for all modern technology.

Friction Regimes

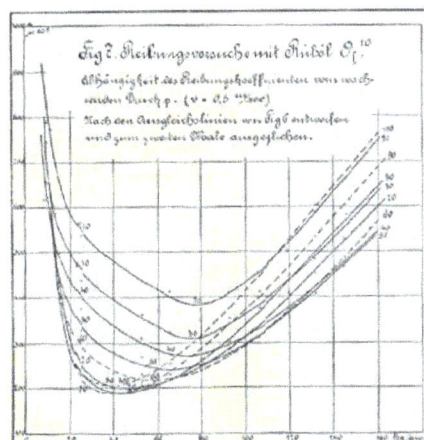

A typical Stribeck curve obtained by Martens

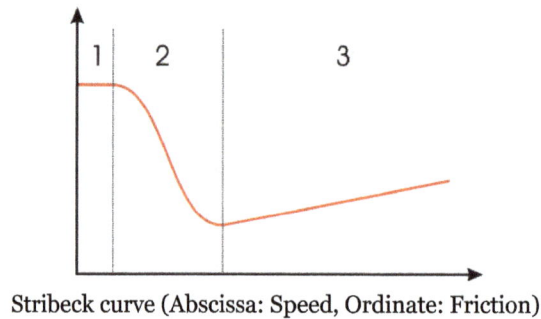

Stribeck curve (Abscissa: Speed, Ordinate: Friction)

1. Solid/boundary friction

2. Mixed friction

3. Fluid friction

Friction regimes for sliding lubricated surfaces have been broadly categorized into:

1.	Solid/boundary friction
2.	Mixed friction
3.	Fluid friction

on the basis of the "Stribeck curve". These curves clearly show the minimum value of friction as the demarcation between full fluid-film lubrication and some solid asperity interactions.

Stribeck and others systematically studied the variation of friction between two liquid lubricated surfaces as a function of a dimensionless lubrication parameter $\eta N/P$, where η is the dynamic viscosity (Ns/m²), N the sliding speed (m/s) and P the load projected on to the geometrical surface (usually load per unit length of bearing in N/m).

The "Stribeck-curve" has been a classic teaching element in tribology classes.

History

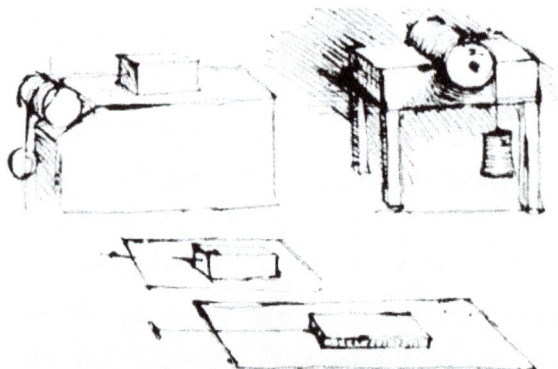

Tribological experiments suggested by Leonardo da Vinci

Duncan Dowson surveyed the history of tribology in his book "History of Tribology (2nd edition)".

This comprehensive book covers developments from prehistory, through early civilizations (Mesopotamia, Egypt) and finally the key developments up to the end of the twentieth century.

Historically, Leonardo da Vinci (1452–1519) was the first to enunciate two laws of friction (it was this connection that gave the name to the Leonardo Centre for Tribology, one of the UK's leading research centres on the subject). According to da Vinci, the frictional resistance was the same for two different objects of the same weight but making contacts over different widths and lengths. He also observed that the force needed to overcome friction doubles when the weight doubles. da Vinci's findings remained unpublished in his notebooks. Da Vinci identified the laws of friction in a notebook in 1493 and continued his studies of friction for 20 years.

Guillaume Amontons rediscovered the classic rules (1699), but unlike da Vinci, made his findings public at the Academie Royale des Sciences for verification. They were further developed by Charles-Augustin de Coulomb (1785).

Charles Hatchett (1760–1820) carried out the first reliable test on frictional wear using a simple reciprocating machine to evaluate wear on gold coins. He found that compared to self-mated coins, coins with grits between them wore at a faster rate.

Stribeck Curve

The "Stribeck curve" or "Stribeck–Hersey curve" (named after Richard Stribeck, who heavily documented and established examples of it, and Mayo D. Hersey), which is used to categorize the friction properties between two surfaces, was developed in the first half of the 20th century. The research of Professor Richard Stribeck (1861–1950) was performed in Berlin at the Royal Prussian Technical Testing Institute (MPA, now BAM). Similar work was previously performed around 1885 by Prof. Adolf Martens (1850–1914) at the same Institute and in the mid-1870s by Dr. Robert H. Thurston at the Stevens Institute of Technology in the U.S. Prof. Dr. Thurston was therefore close to establishing the "Stribeck curve", but he presented no "Stribeck"-like graphs, as he evidently did not fully believe in the relevance of this dependency. Since that time the "Stribeck-curve" has been a classic teaching element in tribology classes.

The graphs of friction force reported by Stribeck stem from a carefully conducted, wide-ranging series of experiments on journal bearings. Stribeck systematically studied the variation of friction between two liquid lubricated surfaces. His results were presented on 5 December 1901 during a public session of the railway society and published on 6 September 1902. They clearly showed the minimum value of friction as the demarcation between full fluid-film lubrication and some solid asperity interactions. Stribeck studied different bearing materials and aspect ratios D/L from 1:1 to 1:2. The maximum sliding speed was 4 m/s and the geometrical contact pressure was limited to 5 MPa.

The reason why the form of the friction curve for liquid lubricated surfaces was later attributed to Stribeck, although both Thurston and Martens achieved their results considerably earlier, (Martens even in the same organization roughly 15 years before), may be because Stribeck published in the most important technical journal in Germany at that time, Zeitschrift des Vereins Deutscher Ingenieure (VDI, Journal of German Mechanical Engineers). Martens published his results "only" in the official journal of the Royal Prussian Technical Testing Institute, which has now become

BAM. The VDI journal, as one of the most important journals for engineers, provided wide access to these data and later colleagues rationalized the results into the three classical friction regimes. Thurston however, did not have the experimental means to record a continuous graph of the coefficient of friction but only measured the friction at discrete points; this may be the reason why the minimum in the coefficient of friction was not discovered by him. Instead, Thurston's data did not indicate such a pronounced minimum of friction for a liquid lubricated journal bearing as was demonstrated by the graphs of Martens and Stribeck.

Jost Report

The term *tribology* became widely used following The Jost Report in 1966. The report said that friction, wear and corrosion were costing the UK huge sums of money every year. As a result, the UK set up several national centres for tribology. Since then the term has diffused into the international engineering field, with many specialists now identifying as tribologists.

There are now numerous national and international societies, such as the *Society for Tribologists and Lubrication Engineers* (STLE) in the USA, the *Institution of Mechanical Engineers' Tribology Group* (IMechE Tribology Group) in the UK or the German Society for Tribology (Gesellschaft für Tribologie, www.gft-ev.de) and MYTRIBOS (Malaysian Tribology society).

Most technical universities have researchers working on tribology, often as part of mechanical engineering departments. The limitations in tribological interactions are, however, no longer mainly determined by mechanical designs, but by material limitations. So the discipline of tribology now counts at least as many materials engineers, physicists and chemists as it does mechanical engineers.

New Areas of Tribology

Since the 1990s, new areas of tribology have emerged, including the nanotribology, biotribology, and green tribology. These interdisciplinary areas study the friction, wear and lubrication at the nanoscale (including the Atomic force microscopy and micro/nanoelectromechanical systems, MEMS/NEMS), in biomedical applications (e.g., human joint prosthetics, dental materials), and ecological aspects of friction, lubrication and wear (tribology of clean energy sources, green lubricants, biomimetic tribology).

Recently, intensive studies of superlubricity (phenomenon of vanishing friction) have sparked due to high demand in energy savings. Development of new materials, such as graphene, initiated development of fundamentally new approaches in the lubrication field.

Applications

The study of tribology is commonly applied in bearing design but extends into almost all other aspects of modern technology, even to such unlikely areas as hair conditioners and cosmetics such as lipstick, powders and lip-gloss.

Any product where one material slides or rubs over another is affected by complex tribological interactions, whether lubricated like hip implants and other artificial prostheses, or unlubricated as in high temperature sliding wear in which conventional lubricants cannot be used but in which the formation of compacted oxide layer glazes have been observed to protect against wear.

Tribology plays an important role in manufacturing. In metal-forming operations, friction increases tool wear and the power required to work a piece. This results in increased costs due to more frequent tool replacement, loss of tolerance as tool dimensions shift, and greater forces required to shape a piece. The use of lubricants which minimize direct surface contact reduces tool wear and power requirements.

References

- Needham, Joseph (1991). Science and Civilisation in China: Volume 2, History of Scientific Thought. Cambridge University Press. ISBN 0-521-05800-7.

- Rosheim, Mark E. (1994). Robot Evolution: The Development of Anthrobotics. Wiley-IEEE. pp. 9–10. ISBN 0-471-02622-0.

- Crane, Carl D.; Joseph Duffy (1998). Kinematic Analysis of Robot Manipulators. Cambridge University Press. ISBN 0-521-57063-8. Retrieved 2007-10-16.

- H. Czichos, K.-H. Habig, Tribologie-Handbuch (Tribology handbook), Vieweg Verlag, Wiesbaden, 2nd edition, 2003, ISBN 3-528-16354-2

- "Focal Points Seminar on review articles in the future of work - Safety and health at work - EU-OSHA". osha.europa.eu. Retrieved 2016-04-19.

- Hutchings, Ian M. (2016-08-15). "Leonardo da Vinci's studies of friction". Wear. 360–361: 51–66. doi:10.1016/j.wear.2016.04.019.

- "Study reveals Leonardo da Vinci's 'irrelevant' scribbles mark the spot where he first recorded the laws of friction". Retrieved 2016-07-26.

- "iSplash-II: Realizing Fast Carangiform Swimming to Outperform a Real Fish" (PDF). Robotics Group at Essex University. Retrieved 2015-09-29.

- "iSplash-I: High Performance Swimming Motion of a Carangiform Robotic Fish with Full-Body Coordination" (PDF). Robotics Group at Essex University. Retrieved 2015-09-29.

- Toy, Tommy (June 29, 2011). "Outlook for robotics and Automation for 2011 and beyond are excellent says expert". PBT Consulting. Retrieved 2012-01-27.

- Lawrence J. Kamm (1996). Understanding Electro-Mechanical Engineering: An Introduction to Mechatronics. John Wiley & Sons. ISBN 978-0-7803-1031-5.

- Swenson, S. Don (1995). HVAC: heating, ventilating, and air conditioning. Homewood, Illinois: American Technical Publishers. ISBN 978-0-8269-0675-5.

- Al-Kodmany, Kheir (2013). The Future of the City: Tall Buildings and Urban Design. WIT Press. "What Does SEER Stand For?". allclimate.net. Retrieved 2015-09-07.

- Pauschinger T. (2012). Solar District Heating with Seasonal Thermal Energy Storage in Germany. European Sustainable Energy Week, Brussels. 18–22 June 2012.

- Holm L. (2012). Long Term Experiences with Solar District Heating in Denmark. European Sustainable Energy Week, Brussels. 18 - 22 June 2012.

- Dianat, Nazari, I,I. "Characteristic of unintentional carbon monoxide poisoning in Northwest Iran- Tabriz". International Journal of Injury Control and Promotion. Retrieved 2011-11-15.

- Mechanical and Mechatronics Engineering Department. "What is Mechatronics Engineering?". Prospective Student Information. University of Waterloo. Retrieved 30 May 2011.

- Faculty of Mechatronics, Informatics and Interdisciplinary Studies TUL. "Mechatronics (Bc., Ing., PhD.)". Retrieved 15 April 2011.

6

Mechanism Engineering: An Integrated Study

The significant aspects of mechanism engineering are discussed in this section. Mechanism is a device designed to transform input forces and movement into a desired set of output forces and movement. The aspects elucidated in this chapter are of vital importance, and provide a better understanding of mechanical engineering.

Mechanism (Engineering)

A mechanism is a device designed to transform input forces and movement into a desired set of output forces and the movement. Mechanisms generally consist of moving components such as gears and gear trains, belt and chain drives, cam and follower mechanisms, and linkages as well as friction devices such as brakes and clutches, and structural components such as the frame, fasteners, bearings, springs, lubricants and seals, as well as a variety of specialized machine elements such as splines, pins and keys.

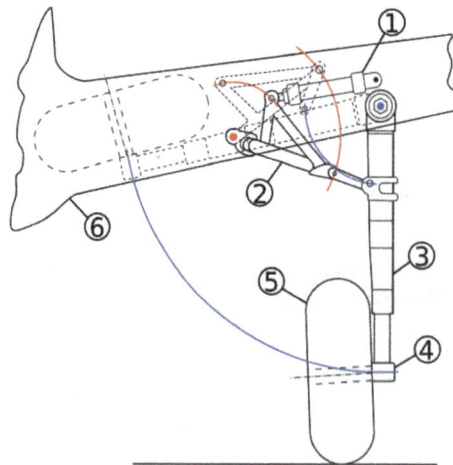

Schematic of the actuator mechanism for an aircraft landing gear.

The German scientist Reuleaux provides the definition "a machine is a combination of resistant bodies so arranged that by their means the mechanical forces of nature can be compelled to do work accompanied by certain determinate motion." In this context, his use of *machine* is generally interpreted to mean *mechanism*.

The combination of force and movement defines power, and a mechanism is designed to manage power in order to achieve a desired set of forces and movement.

A mechanism is usually a piece of a larger process or mechanical system. Sometimes an entire machine may be referred to as a mechanism. Examples are the steering mechanism in a car, or the winding mechanism of a wristwatch. Multiple mechanisms are machines.

Types of Mechanisms

From the time of Archimedes through the Renaissance, mechanisms were considered to be constructed from simple machines, such as the lever, pulley, screw, wheel and axle, wedge and inclined plane. It was Reuleaux who focussed on bodies, called links, and the connections between these bodies called kinematic pairs, or joints.

In order to use geometry to study the movement of a mechanism, its links are modeled as rigid bodies. This means distances between points in a link are assumed to be unchanged as the mechanism moves, that is the link does not flex. Thus, the relative movement between points in two connected links is considered to result from the kinematic pair that joins them.

Kinematic pairs, or joints, are considered to provide ideal constraints between two links, such as the constraint of a single point for pure rotation, or the constraint of a line for pure sliding, as well as pure rolling without slipping and point contact with slipping. A mechanism is modeled as an assembly of rigid links and kinematic pairs.

Kinematic Pairs

Reuleaux called the ideal connections between links kinematic pairs. He distinguished between higher pairs which were said to have line contact between the two links and lower pairs that have area contact between the links. J. Phillips shows that there are many ways to construct pairs that do not fit this simple.

Lower pair: A lower pair is an ideal joint that has surface contact between the pair of elements. We have the following cases:

- A revolute pair, or hinged joint, requires a line in the moving body to remain co-linear with a line in the fixed body, and a plane perpendicular to this line in the moving body maintain contact with a similar perpendicular plane in the fixed body. This imposes five constraints on the relative movement of the links, which therefore has one degree of freedom.

- A prismatic joint, or slider, requires that a line in the moving body remain co-linear with a line in the fixed body, and a plane parallel to this line in the moving body maintain contact with a similar parallel plan in the fixed body. This imposes five constraints on the relative movement of the links, which therefore has one degree of freedom.

- A cylindrical joint requires that a line in the moving body remain co-linear with a line in the fixed body. It is a combination of a revolute joint and a sliding joint. This joint has two degrees of freedom.

- A spherical joint, or ball joint, requires that a point in the moving body maintain contact with a point in the fixed body. This joint has three degrees of freedom.

- A planar joint requires that a plane in the moving body maintain contact with a plane in fixed body. This joint has three degrees of freedom.

- A screw joint, or helical joint, has only one degree of freedom because the sliding and rotational motions are related by the helix angle of the thread.

Higher pairs: Generally, a higher pair is a constraint that requires a line or point contact between the elemental surfaces. For example, the contact between a cam and its follower is a higher pair called a *cam joint*. Similarly, the contact between the involute curves that form the meshing teeth of two gears are cam joints.

Planar Mechanism

A planar mechanism is a mechanical system that is constrained so the trajectories of points in all the bodies of the system lie on planes parallel to a ground plane. The rotational axes of hinged joints that connect the bodies in the system are perpendicular to this ground plane.

Spherical Mechanism

A spherical mechanism is a mechanical system in which the bodies move in a way that the trajectories of points in the system lie on concentric spheres. The rotational axes of hinged joints that connect the bodies in the system pass through the center of these circle .

Spatial Mechanism

A spatial mechanism is a mechanical system that has at least one body that moves in a way that its point trajectories are general space curves. The rotational axes of hinged joints that connect the bodies in the system form lines in space that do not intersect and have distinct common normals.

Gears and Gear Trains

Gears are a type of mechanism.

The transmission of rotation between contacting toothed wheels can be traced back to the Antikythera mechanism of Greece and the south-pointing chariot of China. Illustrations by the renaissance scientist Georgius Agricola show gear trains with cylindrical teeth. The implementation of the involute tooth yielded a standard gear design that provides a constant speed ratio. Some important features of gears and gear trains are:

- The ratio of the pitch circles of mating gears defines the speed ratio and the mechanical advantage of the gear set.

- A planetary gear train provides high gear reduction in a compact package.

- It is possible to design gear teeth for gears that are non-circular, yet still transmit torque smoothly.

- The speed ratios of chain and belt drives are computed in the same way as gear ratios.

Cam and Follower Mechanisms

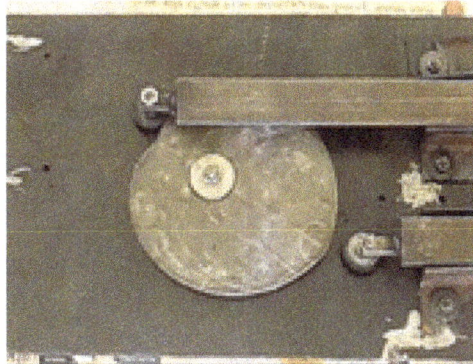

Cam follower Mechanism- Force is Applied From Follower To Cam

A cam and follower is formed by the direct contact of two specially shaped links. The driving link is called the cam and the link that is driven through the direct contact of their surfaces is called the follower. The shape of the contacting surfaces of the cam and follower determines the movement of the mechanism. In general a cam follower mechanism's energy is transferred from cam to follower. The cam shaft is rotated and, according to the cam profile, the follower moves up and down. Now slightly different types of eccentric cam followers are also available in which energy is transferred from the follower to the cam. The main benefit of this type of cam follower mechanism is that the follower moves a little bit and helps to rotate the cam 6 times more circumference length with 70% force.

Linkages

Theo Jansen's kinetic sculpture *Strandbeest*. A wind-driven walking machine.

A linkage is a collection of links connected by joints. Generally, the links are the structural elements and the joints allow movement. Perhaps the single most useful example is the planar four-bar linkage. However, there are many more special linkages:

- Watt's linkage is a four-bar linkage that generates an approximate straight line. It was critical to the operation of his design for the steam engine. This linkage also appears in vehicle suspensions to prevent side-to-side movement of the body relative to the wheels.

- The success of Watt's linkage lead to the design of similar approximate straight-line linkages, such as Hoeken's linkage and Chebyshev's linkage.

- The Peaucellier linkage generates a true straight-line output from a rotary input.

- The Sarrus linkage is a spatial linkage that generates straight-line movement from a rotary input.

- The Klann linkage and the Jansen linkage are recent inventions that provide interesting walking movements. They are respectively a six-bar and an eight-bar linkage.

Flexure Mechanisms

A flexure mechanism consisted of a series of rigid bodies connected by compliant elements (flexure bearings also known as flexure joints) that is designed to produce a geometrically well-defined motion upon application of a force.

Simple Machine

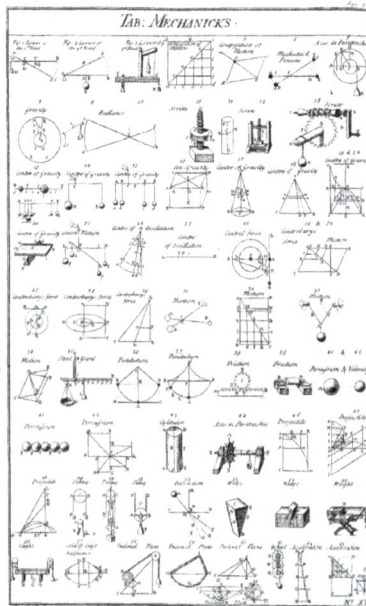

Table of simple mechanisms, from *Chambers' Cyclopædia*, 1728. Simple machines provide a vocabulary for understanding more complex machines.

A simple machine is a mechanical device that changes the direction or magnitude of a force. In general, they can be defined as the simplest mechanisms that use mechanical advantage (also called leverage) to multiply force. Usually the term refers to the six classical simple machines which were defined by Renaissance scientists:

- Lever

- Wheel and axle

- Pulley

- Inclined plane

- Wedge

- Screw

A simple machine uses a single applied force to do work against a single load force. Ignoring friction losses, the work done on the load is equal to the work done by the applied force. The machine can increase the amount of the output force, at the cost of a proportional decrease in the distance moved by the load. The ratio of the output to the applied force is called the *mechanical advantage.*

Simple machines can be regarded as the elementary "building blocks" of which all more complicated machines (sometimes called "compound machines") are composed. For example, wheels, levers, and pulleys are all used in the mechanism of a bicycle. The mechanical advantage of a compound machine is just the product of the mechanical advantages of the simple machines of which it is composed.

Although they continue to be of great importance in mechanics and applied science, modern mechanics has moved beyond the view of the simple machines as the ultimate building blocks of which all machines are composed, which arose in the Renaissance as a neoclassical amplification of ancient Greek texts on technology. The great variety and sophistication of modern machine linkages, which arose during the Industrial Revolution, is inadequately described by these six simple categories. As a result, various post-Renaissance authors have compiled expanded lists of "simple machines", often using terms like *basic machines, compound machines,* or *machine elements* to distinguish them from the classical simple machines above. By the late 1800s, Franz Reuleaux had identified hundreds of machine elements, calling them *simple machines.* Models of these devices may be found at Cornell University's Kinematic Models for Design (KMODDL) website.

History

The idea of a simple machine originated with the Greek philosopher Archimedes around the 3rd century BC, who studied the Archimedean simple machines: lever, pulley, and screw. He discovered the principle of mechanical advantage in the lever. Archimedes' famous remark with regard to the lever: "Give me a place to stand on, and I will move the Earth." expresses his realization that there was no limit to the amount of force amplification that could be achieved by using mechanical advantage. Later Greek philosophers defined the classic five simple machines (excluding the inclined plane) and were able to roughly calculate their mechanical advantage. For example, Heron of Alexandria (ca. 10–75 AD) in his work *Mechanics* lists five mechanisms that can "set a load in motion"; lever, windlass, pulley, wedge, and screw, and describes their fabrication and uses. However the Greeks' understanding was limited to the statics of simple machines; the balance of forces, and did not include dynamics; the tradeoff between force and distance, or the concept of work.

During the Renaissance the dynamics of the *Mechanical Powers*, as the simple machines were called, began to be studied from the standpoint of how far they could lift a load, in addition to the force they could apply, leading eventually to the new concept of mechanical work. In 1586 Flemish engineer Simon Stevin derived the mechanical advantage of the inclined plane, and it was included with the other simple machines. The complete dynamic theory of simple machines was worked out by Italian scientist Galileo Galilei in 1600 in *Le Meccaniche* (*On Mechanics*), in which he showed the underlying mathematical similarity of the machines. He was the first to understand that simple machines do not create energy, only transform it.

The classic rules of sliding friction in machines were discovered by Leonardo da Vinci (1452–1519), but remained unpublished in his notebooks. They were rediscovered by Guillaume Amontons (1699) and were further developed by Charles-Augustin de Coulomb (1785).

Frictionless Analysis

Although each machine works differently mechanically, the way they function is similar mathematically. In each machine, a force F_{in} is applied to the device at one point, and it does work moving a load, F_{out} at another point. Although some machines only change the direction of the force, such as a stationary pulley, most machines multiply the magnitude of the force by a factor, the mechanical advantage

$$\text{MA} = F_{out} / F_{in}$$

that can be calculated from the machine's geometry and friction.

Simple machines do not contain a source of energy, so they cannot do more work than they receive from the input force. A simple machine with no friction or elasticity is called an *ideal machine*. Due to conservation of energy, in an ideal simple machine, the power output (rate of energy output) at any time P_{out} is equal to the power input P_{in}

$$P_{out} = P_{in}$$

The power output equals the velocity of the load P_{in} multiplied by the load force $P_{out} = F_{out} v_{out}$. Similarly the power input from the applied force is equal to the velocity of the input point v_{in} multiplied by the applied force $P_{in} = F_{in} v_{in}$. Therefore,

$$F_{out} v_{out} = F_{in} v_{in}$$

Therefore, the mechanical advantage of a frictionless machine is equal to the *velocity ratio*, the ratio of input velocity to output velocity

$$\boxed{\text{MA}_{ideal} = \frac{F_{out}}{F_{in}} = \frac{v_{in}}{v_{out}}}$$

The *velocity ratio* of the machine is also equal to the ratio of the distance the output point moves to the corresponding distance the input point moves

$$\frac{v_{out}}{v_{in}} = \frac{d_{out}}{d_{in}}$$

This can be calculated from the geometry of the machine. For example, the velocity ratio of the lever is equal to the ratio of its lever arms.

The mechanical advantage can be greater or less than one:

- If $MA > 1$ the output force is greater than the input, the machine acts as a force amplifier, but the distance moved by the load d_{out} is less than the distance moved by the input force d_{in}.

- If $MA < 1$ the output force is less than the input, but the distance moved by the load is greater than the distance moved by the input force.

In the screw, which uses rotational motion, the input force should be replaced by the torque, and the velocity by the angular velocity the shaft is turned.

Friction And Efficiency

All real machines have friction, which causes some of the input power to be dissipated as heat. If P_{fric} is the power lost to friction, from conservation of energy

$$P_{in} = P_{out} + P_{fric}$$

The efficiency η of a machine is a number between 0 and 1 defined as the ratio of power out to the power in, and is a measure of the energy losses

$$\eta \equiv \frac{P_{out}}{P_{in}}$$

$$P_{out} = \eta P_{in}$$

As above, the power is equal to the product of force and velocity, so

$$F_{out} v_{out} = \eta F_{in} v_{in}$$

Therefore,

$$\boxed{MA = \frac{F_{out}}{F_{in}} = \eta \frac{v_{in}}{v_{out}}}$$

So in non-ideal machines, the mechanical advantage is always less than the velocity ratio by the product with the efficiency η. So a machine that includes friction will not be able to move as large a load as a corresponding ideal machine using the same input force.

Compound Machines

A *compound machine* is a machine formed from a set of simple machines connected in series with the output force of one providing the input force to the next. For example, a bench vise consists of a lever (the vise's handle) in series with a screw, and a simple gear train consists of a number of gears (wheels and axles) connected in series.

The mechanical advantage of a compound machine is the ratio of the output force exerted by the last machine in the series divided by the input force applied to the first machine, that is

$$\text{MA}_{\text{compound}} = \frac{F_{\text{outN}}}{F_{\text{in1}}}$$

Because the output force of each machine is the input of the next,
$F_{\text{out1}} = F_{\text{in2}}, F_{\text{out2}} = F_{\text{in3}}, \dots F_{\text{outK}} = F_{\text{inK+1}},$, this mechanical advantage is also given by

$$\text{MA}_{\text{compound}} = \frac{F_{\text{out1}}}{F_{\text{in1}}} \frac{F_{\text{out2}}}{F_{\text{in2}}} \frac{F_{\text{out3}}}{F_{\text{in3}}} \dots \frac{F_{\text{outN}}}{F_{\text{inN}}}$$

Thus, the mechanical advantage of the compound machine is equal to the product of the mechanical advantages of the series of simple machines that form it

$$\text{MA}_{\text{compound}} = \text{MA}_1 \text{MA}_2 \dots \text{MA}_{\text{N}}$$

Similarly, the efficiency of a compound machine is also the product of the efficiencies of the series of simple machines that form it

$$\eta_{\text{compound}} = \eta_1 \eta_2 \dots \eta_{\text{N}}.$$

Self-Locking Machines

The screw's self-locking property is the reason for its wide use in threaded fasteners like bolts and wood screws

In many simple machines, if the load force F_{out} on the machine is high enough in relation to the input force F_{in}, the machine will move backwards, with the load force doing work on the input force. So these machines can be used in either direction, with the driving force applied to either input point. For example, if the load force on a lever is high enough, the lever will move backwards, moving the input arm backwards against the input force. These are called "*reversible*", "*non-locking*" or "*overhauling*" machines, and the backward motion is called "*overhauling*". However, in some machines, if the frictional forces are high enough, no amount of load force can move it backwards, even if the input force is zero. This is called a "*self-locking*", "*nonreversible*", or "*non-overhauling*" machine. These machines can only be set in motion by a force at the input, and when the input force is removed will remain motionless, "locked" by friction at whatever position they were left.

Self-locking occurs mainly in those machines with large areas of sliding contact between moving parts: the screw, inclined plane, and wedge:

The most common example is a screw. In most screws, applying torque to the shaft can cause it to turn, moving the shaft linearly to do work against a load, but no amount of axial load force against the shaft will cause it to turn backwards.

In an inclined plane, a load can be pulled up the plane by a sideways input force, but if the plane is not too steep and there is enough friction between load and plane, when the input force is removed the load will remain motionless and will not slide down the plane, regardless of its weight.

A wedge can be driven into a block of wood by force on the end, such as from hitting it with a sledge hammer, forcing the sides apart, but no amount of compression force from the wood walls will cause it to pop back out of the block.

A machine will be self-locking if and only if its efficiency η is below 50%:

$$\eta \equiv \frac{F_{out}/F_{in}}{d_{in}/d_{out}} < 0.50$$

Whether a machine is self-locking depends on both the friction forces (coefficient of static friction) between its parts, and the distance ratio d_{in}/d_{out} (ideal mechanical advantage). If both the friction and ideal mechanical advantage are high enough, it will self-lock.

Proof

When a machine moves in the forward direction from point 1 to point 2, with the input force doing work on a load force, from conservation of energy the input work $W_{1,2}$ is equal to the sum of the work done on the load force W_{load} and the work lost to friction W_{fric}

$$W_{1,2} = W_{load} + W_{fric} \qquad (1)$$

If the efficiency is below 50% $\eta = W_{load}/W_{1,2} < 1/2$

$$2W_{load} < W_{1,2}$$

From (1)

$$2W_{load} < W_{load} + W_{fric}$$

$$W_{load} < W_{fric}$$

When the machine moves backward from point 2 to point 1 with the load force doing work on the input force, the work lost to friction W_{fric} is the same

$$W_{load} = W_{2,1} + W_{fric}$$

So the output work is

$$W_{2,1} = W_{load} - W_{fric} < 0$$

Thus the machine self-locks, because the work dissipated in friction is greater than the work done by the load force moving it backwards even with no input force

Modern Machine Theory

Kinematic Chains

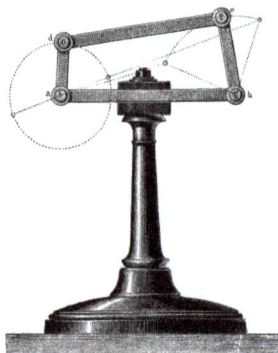

Illustration of a four-bar linkage from Kinematics of Machinery, 1876

Simple machines are elementary examples of kinematic chains that are used to model mechanical systems ranging from the steam engine to robot manipulators. The bearings that form the fulcrum of a lever and that allow the wheel and axle and pulleys to rotate are examples of a kinematic pair called a hinged joint. Similarly, the flat surface of an inclined plane and wedge are examples of the kinematic pair called a sliding joint. The screw is usually identified as its own kinematic pair called a helical joint.

Two levers, or cranks, are combined into a planar four-bar linkage by attaching a link that connects the output of one crank to the input of another. Additional links can be attached to form a six-bar linkage or in series to form a robot.

Classification of Machines

The identification of simple machines arises from a desire for a systematic method to invent new machines. Therefore, an important concern is how simple machines are combined to make more complex machines. One approach is to attach simple machines in series to obtain compound machines.

However, a more successful strategy was identified by Franz Reuleaux, who collected and studied over 800 elementary machines. He realized that a lever, pulley, and wheel and axle are in essence the same device: a body rotating about a hinge. Similarly, an inclined plane, wedge, and screw are a block sliding on a flat surface.

This realization shows that it is the joints, or the connections that provide movement, that are the primary elements of a machine. Starting with four types of joints, the revolute joint, sliding joint, cam joint and gear joint, and related connections such as cables and belts, it is possible to understand a machine as an assembly of solid parts that connect these joints.

Linkage (Mechanical)

A mechanical linkage is an assembly of bodies connected to manage forces and movement. The movement of a body, or link, is studied using geometry so the link is considered to be rigid. The connections between links are modeled as providing ideal movement, pure rotation or sliding for example, and are called joints. A linkage modeled as a network of rigid links and ideal joints is called a kinematic chain.

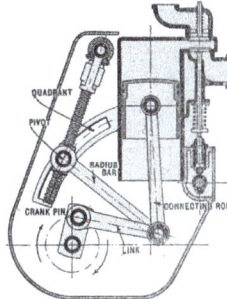

Variable stroke engine (Autocar Handbook, Ninth edition)

Linkages may be constructed from open chains, closed chains, or a combination of open and closed chains. Each link in a chain is connected by a joint to one or more other links. Thus, a kinematic chain can be modeled as a graph in which the links are paths and the joints are vertices, which is called a linkage graph.

The deployable mirror linkage is constructed from a series of rhombus or scissor linkages.

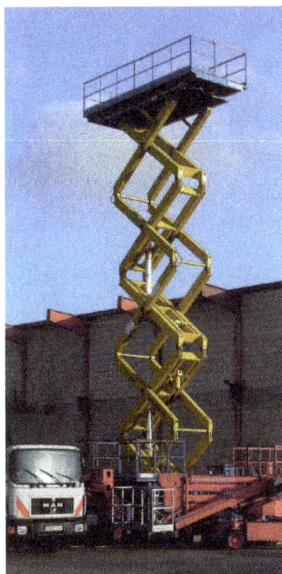

An extended scissor lift

The movement of an ideal joint is generally associated with a subgroup of the group of Euclidean displacements. The number of parameters in the subgroup is called the degrees of freedom (DOF) of the joint. Mechanical linkages are usually designed to transform a given input force and movement into a desired output force and movement. The ratio of the output force to the input force is known as the mechanical advantage of the linkage, while the ratio of the input speed to the output speed is known as the speed ratio. The speed ratio and mechanical advantage are defined so they yield the same number in an ideal linkage.

A kinematic chain, in which one link is fixed or stationary, is called a mechanism, and a linkage designed to be stationary is called a structure.

Uses

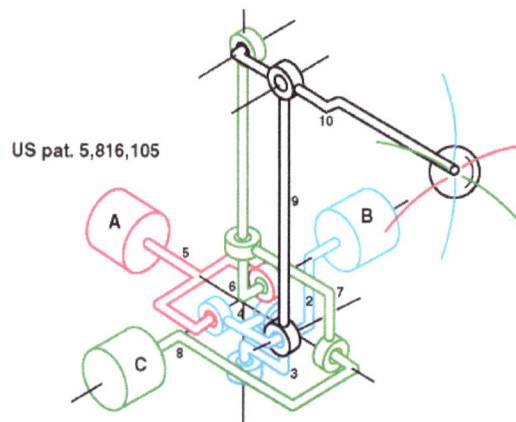

A spatial 3 DOF linkage for joystick applications.

Perhaps the simplest linkage is the lever, which is a link that pivots around a fulcrum attached to ground, or a fixed point. As a force rotates the lever, points far from the fulcrum have a greater velocity than points near the fulcrum. Because power into the lever equals the power out, a small force applied at a point far from the fulcrum (with greater velocity) equals a larger force applied at a point near the fulcrum (with less velocity). The amount the force is amplified is called mechanical advantage. This is the law of the lever.

Two levers connected by a rod so that a force applied to one is transmitted to the second is known as a four-bar linkage. The levers are called cranks, and the fulcrums are called pivots. The connecting rod is also called the coupler. The fourth bar in this assembly is the ground, or frame, on which the cranks are mounted.

Linkages are important components of machines and tools. Examples range from the four-bar linkage used to amplify force in a bolt cutter or to provide independent suspension in an automobile, to complex linkage systems in robotic arms and walking machines. The internal combustion engine uses a slider-crank four-bar linkage formed from its piston, connecting rod, and crankshaft to transform power from expanding burning gases into rotary power. Relatively simple linkages are often used to perform complicated tasks.

Interesting examples of linkages include the windshield wiper, the bicycle suspension, and hydraulic actuators for heavy equipment. In these examples the components in the linkage move in parallel planes and are called "planar linkages." A linkage with at least one link that moves in three-dimensional space is called a "spatial linkage." The skeletons of robotic systems are examples of spatial linkages. The geometric design of these systems relies on modern computer aided design software.

The 4-bar linkage is an adapted mechanical linkage used on bicycles. With a normal full-suspension bike the back wheel moves in a very tight arc shape. This means that more power is lost when going uphill. With a bike fitted with a 4-bar linkage, the wheel moves in such a large arc that it is moving almost vertically. This way the power loss is reduced by up to 30%.

History

Archimedes applied geometry to the study of the lever. Into the 1500s the work of Archimedes and Hero of Alexandria were the primary sources of machine theory. It was Leonardo da Vinci who brought an inventive energy to machines and mechanism.

In the mid-1700s the steam engine was of growing importance, and James Watt realized that efficiency could be increased by using different cylinders for expansion and condensation of the steam. This drove his search for a linkage that could transform rotation of a crank into a linear slide, and resulted in his discovery of what is called Watt's linkage. This led to the study of linkages that could generate straight lines, even if only approximately; and inspired the mathematician J. J. Sylvester, who lectured on the Peaucellier linkage, which generates an exact straight line from a rotating crank.

The work of Sylvester inspired A. B. Kempe, who showed that linkages for addition and multiplication could be assembled into a system that traced a given algebraic curve. Kempe's design procedure has inspired research at the intersection of geometry and computer science.

In the late 1800s F. Reuleaux, A. B. W. Kennedy, and L. Burmester formalized the analysis and synthesis of linkage systems using descriptive geometry, and P.L.Chebyshev introduced analytical techniques for the study and invention of linkages.

In the mid-1900s F. Freudenstein and G. N. Sandor used the newly developed digital computer to solve the loop equations of a linkage and determine its dimensions for a desired function, initiating the computer-aided design of linkages. Within two decades these computer techniques were integral to the analysis of complex machine systems and the control of robot manipulators.

R.E.Kaufman combined the computer's ability to rapidly compute the roots of polynomial equations with a graphical user interface to unite Freudenstein's techniques with the geometrical methods of Reuleaux and Burmester and form *KINSYN,* an interactive computer graphics system for linkage design

The modern study of linkages includes the analysis and design of articulated systems that appear in robots, machine tools, and cable driven and tensegrity systems. These techniques are also being applied to biological systems and even the study of proteins.

Mobility

The configuration of a system of rigid links connected by ideal joints is defined by a set of configuration parameters, such as the angles around a revolute joint and the slides along prismatic joints measured between adjacent links. The geometric constraints of the linkage allow calculation of all of the configuration parameters in terms of a minimum set, which are the *input parameters*. The number of input parameters is called the *mobility*, or degree of freedom, of the linkage system.

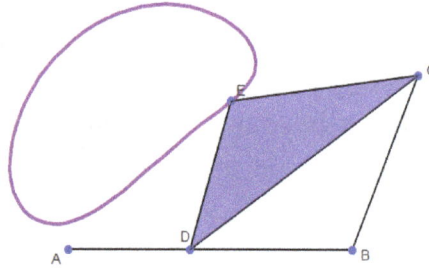

Simple linkages are capable of producing complicated motion.

A system of n rigid bodies moving in space has $6n$ degrees of freedom measured relative to a fixed frame. Include this frame in the count of bodies, so that mobility is independent of the choice of the fixed frame, then we have M=6(N-1), where N=n+1 is the number of moving bodies plus the fixed body.

Joints that connect bodies in this system remove degrees of freedom and reduce mobility. Specifically, hinges and sliders each impose five constraints and therefore remove five degrees of freedom. It is convenient to define the number of constraints c that a joint imposes in terms of the joint's freedom f, where $c=6-f$. In the case of a hinge or slider, which are one degree of freedom joints, we have $f=1$ and therefore $c=6-1=5$.

Thus, the mobility of a linkage system formed from n moving links and j joints each with f_i, $i=1, ...,$ j, degrees of freedom can be computed as,

$$M = 6n - \sum_{i=1}^{j}(6 - f_i) = 6(N-1-j) + \sum_{i=1}^{j} f_i,$$

where N includes the fixed link. This is known as Kutzbach-Gruebler's equation

There are two important special cases: (i) a simple open chain, and (ii) a simple closed chain. A simple open chain consists of n moving links connected end to end by j joints, with one end connected to a ground link. Thus, in this case $N=j+1$ and the mobility of the chain is

$$M = \sum_{i=1}^{j} f_i.$$

For a simple closed chain, n moving links are connected end-to-end by $n+1$ joints such that the two ends are connected to the ground link forming a loop. In this case, we have $N=j$ and the mobility of the chain is

$$M = \sum_{i=1}^{j} f_i - 6.$$

An example of a simple open chain is a serial robot manipulator. These robotic systems are constructed from a series of links connected by six one degree-of-freedom revolute or prismatic joints, so the system has six degrees of freedom.

An example of a simple closed chain is the RSSR spatial four-bar linkage. The sum of the freedom of these joints is eight, so the mobility of the linkage is two, where one of the degrees of freedom is the rotation of the coupler around the line joining the two S joints.

Planar and Spherical Movement

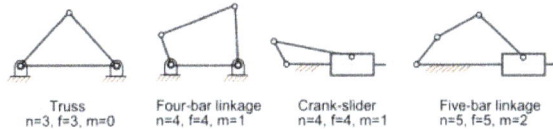

Truss
n=3, f=3, m=0

Four-bar linkage
n=4, f=4, m=1

Crank-slider
n=4, f=4, m=1

Five-bar linkage
n=5, f=5, m=2

Linkage Mobility

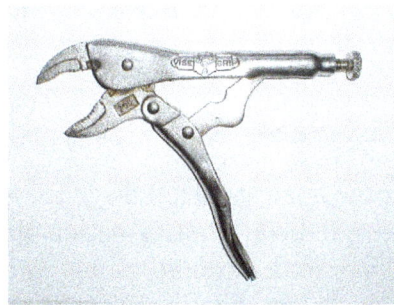

Locking pliers exemplify a four-bar, one degree of freedom mechanical linkage. The adjustable base pivot makes this a two degree-of-freedom five-bar linkage.

It is common practice to design the linkage system so that the movement of all of the bodies are constrained to lie on parallel planes, to form what is known as a *planar linkage*. It is also possible to construct the linkage system so that all of the bodies move on concentric spheres, forming a *spherical linkage*. In both cases, the degrees of freedom of the link is now three rather than six, and the constraints imposed by joints are now $c=3\text{-}f$.

In this case, the mobility formula is given by

$$M = 3(N-1-j) + \sum_{i=1}^{j} f_i,$$

and we have the special cases,

- planar or spherical simple open chain,

$$M = \sum_{i=1}^{j} f_i,$$

- planar or spherical simple closed chain,

$$M = \sum_{i=1}^{j} f_i - 3.$$

An example of a planar simple closed chain is the planar four-bar linkage, which is a four-bar loop with four one degree-of-freedom joints and therefore has mobility M=1.

Joints

The most familiar joints for linkage systems are the revolute, or hinged, joint denoted by an R, and the prismatic, or sliding, joint denoted by a P. Most other joints used for spatial linkages are modeled as combinations of revolute and prismatic joints. For example,

the cylindric joint consists of an RP or PR serial chain constructed so that the axes of the revolute and prismatic joints are parallel,

the spherical joint consists of an RRR serial chain for which each of the hinged joint axes intersect in the same point;

the planar joint can be constructed either as a planar RRR, RPR, and PPR serial chain that has three degrees-of-freedom.

Analysis and Synthesis of Linkages

The primary mathematical tool for the analysis of a linkage is known as the kinematics equations of the system. This is a sequence of rigid body transformation along a serial chain within the linkage that locates a floating link relative to the ground frame. Each serial chain within the linkage that connects this floating link to ground provides a set of equations that must be satisfied by the configuration parameters of the system. The result is a set of non-linear equations that define the configuration parameters of the system for a set of values for the input parameters.

Freudenstein introduced a method to use these equations for the design of a planar four-bar linkage to achieve a specified relation between the input parameters and the configuration of the linkage. Another approach to planar four-bar linkage design was introduced by L. Burmester, and is called Burmester theory.

Planar One Degree-of-Freedom Linkages

The mobility formula provides a way to determine the number of links and joints in a planar linkage that yields a one degree-of-freedom linkage. If we require the mobility of a planar linkage to be M=1 and f_i=1, the result is

$$M = 3(N - 1 - j) + j = 1,$$

or

$$j = (3/2)N - 2.$$

This formula shows that the linkage must have an even number of links, so we have

- N=2, j=1: this is a two-bar linkage known as the lever;

- N=4, j=4: this is the four-bar linkage;

- N=6, j=7: this is a six-bar linkage [it has two links that have three joints, called ternary links, and there are two topologies of this linkage depending how these links are connected. In the Watt topology, the two ternary links are connected by a joint. In the Stephenson topology the two ternary links are connected by binary links;

- N=8, j=10: the eight-bar linkage has 16 different topologies;

- N=10, j=13: the 10-bar linkage has 230 different topologies,

- N=12, j=16: the 12-bar has 6856 topologies.

Sunkari and Schmidt for the number of 14- and 16-bar topologies, as well as the number of linkages that have two, three and four degrees-of-freedom.

The planar four-bar linkage is probably the simplest and most common linkage. It is a one degree-of-freedom system that transforms an input crank rotation or slider displacement into an output rotation or slide.

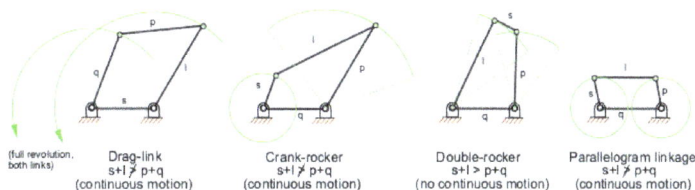

Types of four-bar linkages, s = shortest link, l = longest link

Examples of four-bar linkages are:

- the crank-rocker, in which the input crank fully rotates and the output link rocks back and forth;

- the slider-crank, in which the input crank rotates and the output slide moves back and forth;

- drag-link mechanisms, in which the input crank fully rotates and drags the output crank in a fully rotational movement.

Other Interesting Linkages

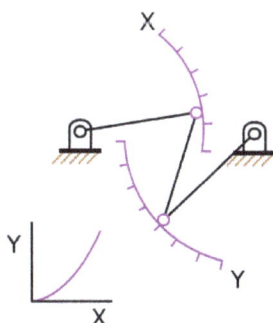

A function generator linkage that approximates a parabolic output.

- Pantograph (four-bar, two DOF)

- Five bar linkages often have meshing gears for two of the links, creating a one DOF linkage. They can provide greater power transmission with more design flexibility than four-bar linkages.

- Jansen's linkage is a twelve-bar Leg mechanism that was invented by kinetic sculptor Theo Jansen.

- Klann linkage is a six-bar linkage that forms the leg of a walking mechanism;

- Toggle mechanisms are four-bar linkages that are dimensioned so that they can fold and lock. The toggle positions are determined by the colinearity of two of the moving links. The linkage is dimensioned so that the linkage reaches a toggle position just before it folds. The high mechanical advantage allows the input crank to deform the linkage just enough to push it beyond the toggle position. This locks the input in place. Toggle mechanisms are used as clamps.

Straight Line Mechanisms

- James Watt's parallel motion and Watt's linkage

- Peaucellier–Lipkin linkage, the first planar linkage to create a perfect straight line output from rotary input; eight-bar, one DOF.

- A Scott Russell linkage, which converts linear motion, to (almost) linear motion in a line perpendicular to the input.

- Chebyshev linkage, which provides nearly straight motion of a point with a four-bar linkage.

- Hoekens linkage, which provides nearly straight motion of a point with a four-bar linkage.

- Sarrus linkage, which provides motion of one surface in a direction normal to another.

- Hart's inversor, which provides a perfect straight line motion without sliding guides.

Biological Linkages

Linkage systems are widely distributed in animals. The most thorough overview of the different types of linkages in animals has been provided by Mees Muller, who also designed a new classification system which is especially well suited for biological systems. A well-known example is the cruciate ligaments of the knee.

An important difference between biological and engineering linkages is that revolving bars are rare in biology and that usually only a small range of the theoretically possible is possible due to additional mechanical constraints (especially the necessity to deliver blood). Biological linkages frequently are compliant. Often one or more bars are formed by ligaments, and often the linkages are three-dimensional. Coupled linkage systems are known, as well as five-, six-, and even seven-bar linkages. Four-bar linkages are by far the most common though.

Linkages can be found in joints, such as the knee of tetrapods, the hock of sheep, and the cranial mechanism of birds and reptiles. The latter is responsible for the upward motion of the upper bill in many birds.

Linkage mechanisms are especially frequent and manifold in the head of bony fishes, such as wrasses, which have evolved many specialized feeding mechanisms. Especially advanced are the linkage mechanisms of jaw protrusion. For suction feeding a system of linked four-bar linkages is responsible for the coordinated opening of the mouth and 3-D expansion of the buccal cavity. Other linkages are responsible for protrusion of the premaxilla.

Linkages are also present as locking mechanisms, such as in the knee of the horse, which enables the animal to sleep standing, without active muscle contraction. In pivot feeding, used by certain bony fishes, a four-bar linkage at first locks the head in a ventrally bent position by the alignment of two bars. The release of the locking mechanism jets the head up and moves the mouth toward the prey within 5-10 ms.

Cam

Fig. 1 Animation showing rotating cams and cam followers producing reciprocating motion.

A cam is a rotating or sliding piece in a mechanical linkage used especially in transforming rotary motion into linear motion or vice versa. It is often a part of a rotating wheel (e.g. an eccentric wheel) or shaft (e.g. a cylinder with an irregular shape) that strikes a lever at one or more points on its circular path. The cam can be a simple tooth, as is used to deliver pulses of power to a steam hammer, for example, or an eccentric disc or other shape that produces a smooth reciprocating (back and forth) motion in the *follower*, which is a lever making contact with the cam.

Overview

The cam can be seen as a device that rotates from circular to reciprocating (or sometimes oscillating) motion. A common example is the camshaft of an automobile, which takes the rotary motion of the engine and translates it into the reciprocating motion necessary to operate the intake and exhaust valves of the cylinders.

Displacement Diagram

Certain cams can be characterized by their displacement diagrams, which reflect the changing position a roller follower (a shaft with a rotating wheel at the end) would make as the cam rotates about an axis. These diagrams relate angular position, usually in degrees, to the radial displacement experienced at that position. Displacement diagrams are traditionally presented as graphs with non-negative values. A simple displacement diagram illustrates the follower motion at a constant velocity rise followed by a similar return with a dwell in between as depicted in figure 2. The rise is the motion of the follower away from the cam center, dwell is the motion where the follower is at rest, and return is the motion of the follower toward the cam center.

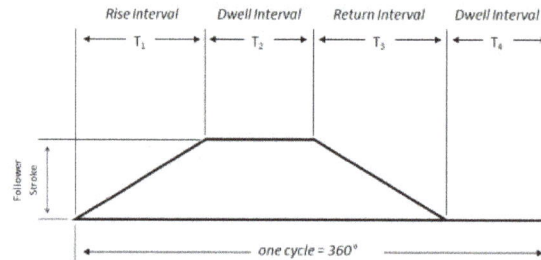

Fig. 2 Basic displacement diagram

However, the most common type is in the valve actuators in internal combustion engines. Here, the cam profile is commonly symmetric and at rotational speeds generally met with, very high acceleration forces develop. Ideally, a convex curve between the onset and maximum position of lift reduces acceleration, but this requires impractically large shaft diameters relative to lift. Thus, in practice, the points at which lift begins and ends mean that a tangent to the base circle appears on the profile. This is continuous with a tangent to the tip circle. In designing the cam, the lift and the dwell angle θ are given. If the profile is treated as a large base circle and a small tip circle, joined by a common tangent, giving lift L, the relationship can be calculated, given the angle ϕ between one tangent and the axis of symmetry (ϕ being $\pi/2 - \theta/2$), while   is the distance between the centres of the circles (required), and R is the radius of the base (given) and r that of the tip circle (required)

$$C = L/(1 - \sin\phi) \text{ and } r = R - L\sin\phi/(1 - \sin\phi)$$

Plate Cam

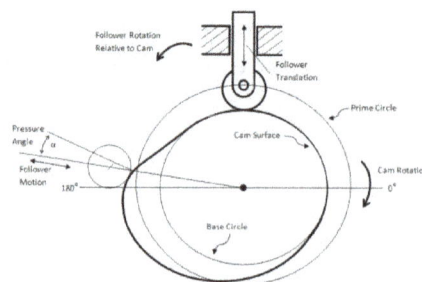

Fig. 3 Cam profile

The most commonly used cam is the plate cam (also *disc cam* or *radial cam*) which is cut out of a piece of flat metal or plate. Here, the follower moves in a plane perpendicular to the axis of rotation of the camshaft. Several key terms are relevant in such a construction of plate cams: base circle, prime circle (with radius equal to the sum of the follower radius and the base circle radius), pitch curve which is the radial curve traced out by applying the radial displacements away from the prime circle across all angles, and the lobe separation angle (LSA - the angle between two adjacent intake and exhaust cam lobes).

The base circle is the smallest circle that can be drawn to the cam profile.

A once common, but now outdated, application of this type of cam was automatic machine tool programming cams. Each tool movement or operation was controlled directly by one or more cams. Instructions for producing programming cams and cam generation data for the most common makes of machine were included in engineering references well into the modern CNC era.

This type of cam is used in many simple electromechanical appliance controllers, such as dishwashers and clothes washing machines, to actuate mechanical switches that control the various parts.

Cylindrical Cam

Motorcycle transmission showing cylindrical cam with three followers. Each follower controls the position of a shift fork.

Constant lead barrel cam in an American Pacemaker lathe. This cam is used to provide a repeatable cross slide setting when threading with a single-point tool.

A cylindrical cam or barrel cam is a cam in which the follower rides on the surface of a cylinder. In the most common type, the follower rides in a groove cut into the surface of a cylinder. These cams are principally used to convert rotational motion to linear motion parallel to the rotational axis of the cylinder. A cylinder may have several grooves cut into the surface and drive several followers. Cylindrical cams can provide motions that involve more than a single rotation of the cylinder and generally provide positive positioning, removing the need for a spring or other provision to keep the follower in contact with the control surface.

Applications include machine tool drives, such as reciprocating saws, and shift control barrels in sequential transmissions, such as on most modern motorcycles.

A special case of this cam is constant lead, where the position of the follower is linear with rotation, as in a lead screw. The purpose and detail of implementation influence whether this application is called a cam or a screw thread, but in some cases, the nomenclature may be ambiguous.

Cylindrical cams may also be used to reference an output to two inputs, where one input is rotation of the cylinder, and the second is position of the follower axially along the cam. The output is radial to the cylinder. These were once common for special functions in control systems, such as fire control mechanisms for guns on naval vessels and mechanical analog computers.

An example of a cylindrical cam with two inputs is provided by a duplicating lathe, an example of which is the Klotz axe handle lathe, which cuts an axe handle to a form controlled by a pattern acting as a cam for the lathe mechanism.

Face Cam

A face cam produces motion by using a follower riding on the face of a disk. The most common type has the follower ride in a slot so that the captive follower produces radial motion with positive positioning without the need for a spring or other mechanism to keep the follower in contact with the control surface. A face cam of this type generally has only one slot for a follower on each face. In some applications, a single element, such as a gear, a barrel cam, or other rotating element with a flat face, may do duty as a face cam in addition to other purposes.

Face cams may provide repetitive motion with a groove that forms a closed curve, or may provide function generation with a stopped groove. Cams used for function generation may have grooves that require several revolutions to cover the complete function, and in this case, the function generally needs to be invertible so that the groove does not self intersect, and the function output value must differ enough at corresponding rotations that there is sufficient material separating the adjacent groove segments. A common form is the constant lead cam, where displacement of the follower is linear with rotation, such as the scroll plate in a scroll chuck. Non-invertible functions, which require the groove to self-intersect, can be implemented using special follower designs.

Sash window lock, traditional cam style, for double-hung sash window

A variant of the face cam provides motion parallel to the axis of cam rotation. A common example is the traditional sash window lock, where the cam is mounted to the top of the lower sash, and the follower is the hook on the upper sash. In this application, the cam is used to provide mechanical advantage in forcing the window shut, and also provides a self-locking action, like some worm gears, due to friction.

Face cams may also be used to reference a single output to two inputs, typically where one input is rotation of the cam and the other is radial position of the follower. The output is parallel to the axis of the cam. These were once common is mechanical analog computation and special functions in control systems.

A face cam that implements three outputs for a single rotational input is the stereo phonograph, where a relatively constant lead groove guides the stylus and tone arm unit, acting as either a rocker-type (tone arm) or linear (linear tracking turntable) follower, and the stylus alone acting as the follower for two orthogonal outputs to representing the audio signals. These motions are in a plane radial to the rotation of the record and at angles of 45 degrees to the plane of the disk (normal to the groove faces). The position of the tone arm was used by some turntables as a control input, such as to turn the unit off or to load the next disk in a stack, but was ignored in simple units.

Heart Shaped Cam

This type of cam, in the form of a symmetric heart symbol, is used to return a shaft holding the cam to a set position by pressure from a roller. They were used for example on early models of Post Office Master clocks to synchronise the clock time with Greenwich Mean Time when the activating follower was pressed onto the cam automatically via a signal from an accurate time source.

Snail Drop Cam

This type of cam was used for example in mechanical time keeping clocking-in clocks to drive the day advance mechanism at precisely midnight and consisted of a follower being raised over 24 hours by the cam in a spiral path which terminated at a sharp cut off at which the follower would drop down and activate the day advance. Where timing accuracy is required as in clocking-in clocks these were typically ingeniously arranged to have a roller cam follower to raise the drop weight for most of its journey to near its full height, and only for the last portion of its travel for the weight to be taken over and supported by a solid follower with a sharp edge. This ensured that the weight dropped at a precise moment, enabling accurate timing. This was achieved by the use of two snail cams mounted coaxially with the roller initially resting on one cam and the final solid follower on the other but not in contact with its cam profile. Thus the roller cam was initially carried the weight, until at the final portion of the run the profile of the non-roller cam rose more than the other causing the solid follower to take the weight.

Linear Cam

A linear cam is one in which the cam element moves in a straight line rather than rotates. The cam element is often a plate or block, but may be any cross section. The key feature is that the input is a linear motion rather than rotational. The cam profile may be cut into one or more edges of a plate or block, may be one or more slots or grooves in the face of an element, or may even be a surface profile for a cam with more than one input. The development of a linear cam is similar to, but not identical to, that of a rotating cam.

A common example of a linear cam is a key for a pin tumbler lock. The pins act as the followers. This behavior is exemplified when the key is duplicated in a key duplication machine, where the original key acts as a control cam for cutting the new key.

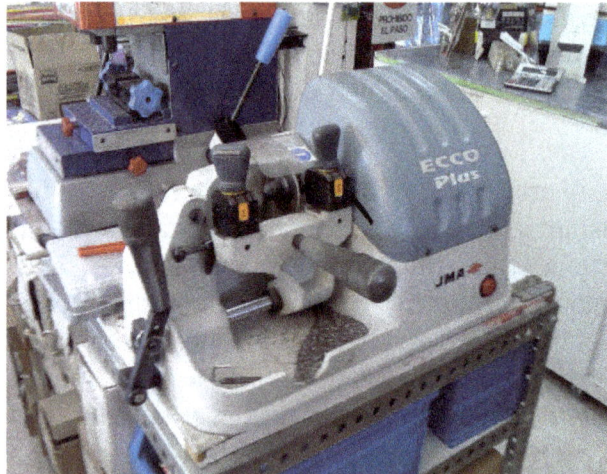

Key duplicating machine. The original key (mounted in the left hand holder) acts as a linear cam to control the cut depth for the duplicate.

History

An early cam was built into Hellenistic water-driven automata from the 3rd century BC. Cams were later employed by Al-Jazari, who used them in his own automata. The cam and camshaft appeared in European mechanisms from the 14th century.

Gear Train

A gear train is a mechanical system formed by mounting gears on a frame so that the teeth of the gears engage.

Gear teeth are designed to ensure the pitch circles of engaging gears roll on each other without slipping, providing a smooth transmission of rotation from one gear to the next.

The transmission of rotation between contacting toothed wheels can be traced back to the Antikythera mechanism of Greece and the south-pointing chariot of China. Illustrations by the Renaissance scientist Georgius Agricola show gear trains with cylindrical teeth. The implementation of the involute tooth yielded a standard gear design that provides a constant speed ratio.

Features of gears and gear trains include:

- The ratio of the pitch circles of mating gears defines the speed ratio and the mechanical advantage of the gear set.

- A planetary gear train provides high gear reduction in a compact package.

- It is possible to design gear teeth for gears that are non-circular, yet still transmit torque smoothly.

- The speed ratios of chain and belt drives are computed in the same way as gear ratios.

An Agricola illustration from 1580 showing a toothed wheel that engages a slotted cylinder to form a gear train that transmits power from a human-powered treadmill to mining pump.

Mechanical Advantage

Gear teeth are designed so that the number of teeth on a gear is proportional to the radius of its pitch circle, and so that the pitch circles of meshing gears roll on each other without slipping. The speed ratio for a pair of meshing gears can be computed from ratio of the radii of the pitch circles and the ratio of the number of teeth on each gear.

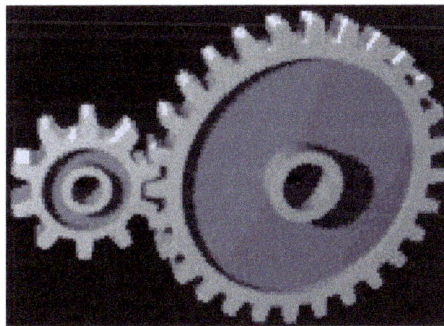

Two meshing gears transmit rotational motion.

The velocity v of the point of contact on the pitch circles is the same on both gears, and is given by

$$v = r_A \omega_A = r_B \omega_B,$$

where input gear A with radius r_A and angular velocity ω_A meshes with output gear B with radius r_B and angular velocity ω_B. Therefore,

$$\frac{\omega_A}{\omega_B} = \frac{r_B}{r_A} = \frac{N_B}{N_A}.$$

where N_A is the number of teeth on the input gear and N_B is the number of teeth on the output gear.

The mechanical advantage of a pair of meshing gears for which the input gear has N_A teeth and the output gear has N_B teeth is given by

$$MA = \frac{T_B}{T_A} = \frac{N_B}{N_A}.$$

This shows that if the output gear G_B has more teeth than the input gear G_A, then the gear train *amplifies* the input torque. And, if the output gear has fewer teeth than the input gear, then the gear train *reduces* the input torque.

If the output gear of a gear train rotates more slowly than the input gear, then the gear train is called a *speed reducer*. In this case, because the output gear must have more teeth than the input gear, the speed reducer amplifies the input torque.

Analysis Using Virtual Work

For this analysis, we consider a gear train that has one degree-of-freedom, which means the angular rotation of all the gears in the gear train are defined by the angle of the input gear.

The size of the gears and the sequence in which they engage define the ratio of the angular velocity ω_A of the input gear to the angular velocity ω_B of the output gear, known as the speed ratio, or gear ratio, of the gear train. Let R be the speed ratio, then

$$\frac{\omega_A}{\omega_B} = R.$$

The input torque T_A acting on the input gear G_A is transformed by the gear train into the output torque T_B exerted by the output gear G_B. If we assume, that the gears are rigid and that there are no losses in the engagement of the gear teeth, then the principle of virtual work can be used to analyze the static equilibrium of the gear train.

Let the angle θ of the input gear be the generalized coordinate of the gear train, then the speed ratio R of the gear train defines the angular velocity of the output gear in terms of the input gear, that is

$$\omega_A = \omega, \quad \omega_B = \omega / R.$$

The formula for the generalized force obtained from the principle of virtual work with applied torques yields

$$F_\theta = T_A \frac{\partial \omega_A}{\partial \omega} - T_B \frac{\partial \omega_B}{\partial \omega} = T_A - T_B / R = 0.$$

The mechanical advantage of the gear train is the ratio of the output torque T_B to the input torque T_A, and the above equation yields

$$MA = \frac{T_B}{T_A} = R.$$

Thus, the speed ratio of a gear train also defines its mechanical advantage. This shows that if the input gear rotates faster than the output gear, then the gear train amplifies the input torque. And, if the input gear rotates slower than the output gear, then the gear train reduces the input torque.

Gear Trains With Two Gears

The simplest example of a gear train has two gears. The "input gear" (also known as drive gear) transmits power to the "output gear" (also known as driven gear). The input gear will typically be connected to a power source, such as a motor or engine. In such an example, the power output of the output (driven) gear depends on the ratio of the dimensions of the two gears.

Formula

The teeth on gears are designed so that the gears can roll on each other smoothly (without slipping or jamming). In order for two gears to roll on each other smoothly, they must be designed so that the velocity at the point of contact of the two pitch circles (represented by v) is the same for each gear.

Mathematically, if the input gear G_A has the radius r_A and angular velocity ω_A, and meshes with output gear G_B of radius r_B and angular velocity ω_B, then:

$$v = r_A \omega_A = r_B \omega_B,$$

The number of teeth on a gear is proportional to the radius of its pitch circle, which means that the ratios of the gears' angular velocities, radii, and number of teeth are equal. Where N_A is the number of teeth on the input gear and N_B is the number of teeth on the output gear, the following equation is formed:

$$\frac{\omega_A}{\omega_B} = \frac{r_B}{r_A} = \frac{N_B}{N_A}.$$

This shows that a simple gear train with two gears has the gear ratio R given by

$$R = \frac{\omega_A}{\omega_B} = \frac{N_B}{N_A}.$$

This equation shows that if the number of teeth on the output gear G_B is larger than the number of teeth on the input gear G_A, then the input gear G_A must rotate faster than the output gear G_B.

Speed Ratio

Gear teeth are distributed along the circumference of the pitch circle so that the thickness t of each tooth and the space between neighboring teeth are the same. The pitch p of a gear, which is the distance between equivalent points on neighboring teeth along the pitch circle, is equal to twice the thickness of a tooth,

$$p = 2t.$$

The pitch of a gear G_A can be computed from the number of teeth N_A and the radius r_A of its pitch circle

$$p = \frac{2\pi r_A}{N_A}.$$

In order to mesh smoothly two gears G_A and G_B must have the same sized teeth and therefore they must have the same pitch p, which means

$$p = \frac{2\pi r_A}{N_A} = \frac{2\pi r_B}{N_B}.$$

This equation shows that the ratio of the circumference, the diameters and the radii of two meshing gears is equal to the ratio of their number of teeth,

$$\frac{r_B}{r_A} = \frac{N_B}{N_A}.$$

The speed ratio of two gears rolling without slipping on their pitch circles is given by,

$$R = \frac{\omega_A}{\omega_B} = \frac{r_B}{r_A},$$

therefore

$$R = \frac{\omega_A}{\omega_B} = \frac{N_B}{N_A}.$$

In other words, the gear ratio, or speed ratio, is inversely proportional to the radius of the pitch circle and the number of teeth of the input gear.

Torque Ratio

A gear train can be analyzed using the principle of virtual work to show that its torque ratio, which is the ratio of its output torque to its input torque, is equal to the gear ratio, or speed ratio, of the gear train.

This means that the input torque T_A applied to the input gear G_A and the output torque T_B on the output gear G_B are related by the ratio

$$R = \frac{T_B}{T_A},$$

where R is the gear ratio of the gear train.

The torque ratio of a gear train is also known as its mechanical advantage

$$MA = \frac{T_B}{T_A}.$$

Idler Gears

In a sequence of gears chained together, the ratio depends only on the number of teeth on the first and last gear. The intermediate gears, regardless of their size, do not alter the overall gear ratio of the chain. However, the addition of each intermediate gear reverses the direction of rotation of the final gear.

An intermediate gear which does not drive a shaft to perform any work is called an idler gear. Sometimes, a single idler gear is used to reverse the direction, in which case it may be referred to as a *reverse idler*. For instance, the typical automobile manual transmission engages reverse gear by means of inserting a reverse idler between two gears.

Idler gears can also transmit rotation among distant shafts in situations where it would be impractical to simply make the distant gears larger to bring them together. Not only do larger gears occupy more space, the mass and rotational inertia (moment of inertia) of a gear is proportional to the square of its radius. Instead of idler gears, a toothed belt or chain can be used to transmit torque over distance.

Formula

If a simple gear train has three gears, such that the input gear G_A meshes with an intermediate gear G_I which in turn meshes with the output gear G_B, then the pitch circle of the intermediate gear rolls without slipping on both the pitch circles of the input and output gears. This yields the two relations

$$\frac{\omega_A}{\omega_I} = \frac{N_I}{N_A}, \quad \frac{\omega_I}{\omega_B} = \frac{N_B}{N_I}.$$

The speed ratio of this gear train is obtained by multiplying these two equations to obtain

$$R = \frac{\omega_A}{\omega_B} = \frac{N_B}{N_A}.$$

Notice that this gear ratio is exactly the same as for the case when the gears G_A and G_B engaged directly. The intermediate gear provides spacing but does not affect the gear ratio. For this reason it is called an *idler* gear. The same gear ratio is obtained for a sequence of idler gears and hence an idler gear is used to provide the same direction to rotate the driver and driven gear, if the driver gear moves in clockwise direction, then the driven gear also moves in the clockwise direction with the help of the idler gear.

Example

2 gears and an idler gear on a piece of farm equipment, with a ratio of 42/13 = 3.23:1

In the photo, assuming that the smallest gear is connected to the motor, it is called the drive gear or input gear. The somewhat larger gear in the middle is called an idler gear. It is not connected directly to either the motor or the output shaft and only transmits power between the input and output gears. There is a third gear in the upper-right corner of the photo. Assuming that that gear is connected to the machine's output shaft, it is the output or driven gear.

The input gear in this gear train has 13 teeth and the idler gear has 21 teeth. Considering only these gears, the gear ratio between the idler and the input gear can be calculated as if the idler gear was the output gear. Therefore, the gear ratio is driven/drive = 21/13 ≈1.62 or 1.62:1.

At this ratio it means that the drive gear must make 1.62 revolutions to turn the driven gear once. It also means that for every one revolution of the driver, the driven gear has made 1/1.62, or 0.62, revolutions. Essentially, the larger gear turns slower.

The third gear in the picture has 42 teeth. The gear ratio between the idler and third gear is thus 42/21, or 2:1, and hence the final gear ratio is 1.62x2≈3.23. For every 3.23 revolutions of the smallest gear, the largest gear turns one revolution, or for every one revolution of the smallest gear, the largest gear turns 0.31 (1/3.23) revolution, a total reduction of about 1:3.23 (Gear Reduction Ratio (GRR) = 1/Gear Ratio (GR)).

Since the idler gear contacts directly both the smaller and the larger gear, it can be removed from the calculation, also giving a ratio of 42/13≈3.23. The idler gear serves to make both the drive gear and the driven gear rotate in the same direction, but confers no mechanical advantage.

Belt Drives

Belts can have teeth in them also and be coupled to gear-like pulleys. Special gears called sprockets can be coupled together with chains, as on bicycles and some motorcycles. Again, exact accounting of teeth and revolutions can be applied with these machines.

Valve timing gears on a Ford Taunus V4 engine — the small gear is on the crankshaft, the larger gear is on the camshaft. The crankshaft gear has 34 teeth, the camshaft gear has 68 teeth and runs at half the crankshaft RPM.(The small gear in the lower left is on the balance shaft.)

For example, a belt with teeth, called the timing belt, is used in some internal combustion engines to synchronize the movement of the camshaft with that of the crankshaft, so that the valves open and close at the top of each cylinder at exactly the right time relative to the movement of each piston. A chain, called a timing chain, is used on some automobiles for this purpose, while in others, the camshaft and crankshaft are coupled directly together through meshed gears. Regardless of which form of drive is employed, the crankshaft to camshaft gear ratio is always 2:1 on four-stroke engines, which means that for every two revolutions of the crankshaft the camshaft will rotate once.

Automotive Applications

Illustration of gears of an automotive transmission

Automobile drivetrains generally have two or more major areas where gearing is used. Gearing is employed in the transmission, which contains a number of different sets of gears that can be changed to allow a wide range of vehicle speeds, and also in the differential, which contains the final drive to provide further speed reduction at the wheels. In addition, the differential contains further gearing that splits torque equally between the two wheels while permitting them to have different speeds when travelling in a curved path. The transmission and final drive might be separate and connected by a driveshaft, or they might be combined into one unit called a transaxle. The gear ratios in transmission and final drive are important because different gear ratios will change the characteristics of a vehicle's performance.

Example

A 2004 Chevrolet Corvette C5 Z06 with a six-speed manual transmission has the following gear ratios in the transmission:

Gear	Ratio
1st gear	2.97:1
2nd gear	2.07:1
3rd gear	1.43:1
4th gear	1.00:1
5th gear	0.84:1
6th gear	0.56:1
reverse	-3.38:1

In 1st gear, the engine makes 2.97 revolutions for every revolution of the transmission's output. In 4th gear, the gear ratio of 1:1 means that the engine and the transmission's output rotate at the same speed. 5th and 6th gears are known as overdrive gears, in which the output of the transmission is revolving faster than the engine's output.

The Corvette above has an axle ratio of 3.42:1, meaning that for every 3.42 revolutions of the transmission's output, the wheels make one revolution. The differential ratio multiplies with the transmission ratio, so in 1st gear, the engine makes 10.16 revolutions for every revolution of the wheels.

The car's tires can almost be thought of as a third type of gearing. This car is equipped with 295/35-18 tires, which have a circumference of 82.1 inches. This means that for every complete revolution of the wheel, the car travels 82.1 inches (209 cm). If the Corvette had larger tires, it would travel farther with each revolution of the wheel, which would be like a higher gear. If the car had smaller tires, it would be like a lower gear.

With the gear ratios of the transmission and differential, and the size of the tires, it becomes possible to calculate the speed of the car for a particular gear at a particular engine RPM.

For example, it is possible to determine the distance the car will travel for one revolution of the engine by dividing the circumference of the tire by the combined gear ratio of the transmission and differential.

$$d = \frac{c_t}{gr_t \times gr_d}$$

It is also possible to determine a car's speed from the engine speed by multiplying the circumference of the tire by the engine speed and dividing by the combined gear ratio.

$$v_c = \frac{c_t \times v_e}{gr_t \times gr_d}$$

Gear	Distance per engine revolution	Speed per 1000 RPM
1st gear	8.1 in (210 mm)	7.7 mph (12.4 km/h)
2nd gear	11.6 in (290 mm)	11.0 mph (17.7 km/h)
3rd gear	16.8 in (430 mm)	15.9 mph (25.6 km/h)
4th gear	24.0 in (610 mm)	22.7 mph (36.5 km/h)
5th gear	28.6 in (730 mm)	27.1 mph (43.6 km/h)
6th gear	42.9 in (1,090 mm)	40.6 mph (65.3 km/h)

Wide-Ratio Vs. Close-Ratio Transmission

A close-ratio transmission is a transmission in which there is a relatively little difference between the gear ratios of the gears. For example, a transmission with an engine shaft to drive shaft ratio of 4:1 in first gear and 2:1 in second gear would be considered wide-ratio when compared to another transmission with a ratio of 4:1 in first and 3:1 in second. This is because the close-ratio transmission has less of a progression between gears. For the wide-ratio transmission, the first gear ratio is 4:1 or 4, and in second gear it is 2:1 or 2, so the progression is equal to 4/2 = 2 (or 200%). For the close-ratio transmission, first gear has a 4:1 ratio or 4, and second gear has a ratio of 3:1 or 3, so the progression between gears is 4/3, or 133%. Since 133% is less than 200%, the transmission with the smaller progression between gears is considered close-ratio. However, the difference between a close-ratio and wide-ratio transmission is subjective and relative.

Close-ratio transmissions are generally offered in sports cars, sport bikes, and especially in race vehicles, where the engine is tuned for maximum power in a narrow range of operating speeds, and the driver or rider can be expected to shift often to keep the engine in its power band.

Factory 4-speed or 5-speed transmission ratios generally have a greater difference between gear ratios and tend to be effective for ordinary driving and moderate performance use. Wider gaps between ratios allow a higher 1st gear ratio for better manners in traffic, but cause engine speed to decrease more when shifting. Narrowing the gaps will increase acceleration at speed, and potentially improve top speed under certain conditions, but acceleration from a stopped position and operation in daily driving will suffer.

Range is the torque multiplication difference between 1st and 4th gears; wider-ratio gear-sets have more, typically between 2.8 and 3.2. This is the single most important determinant of low-speed acceleration from stopped.

Progression is the reduction or decay in the percentage drop in engine speed in the next gear, for example after shifting from 1st to 2nd gear. Most transmissions have some degree of progression in that the RPM drop on the 1-2 shift is larger than the RPM drop on the 2-3 shift, which is in turn larger than the RPM drop on the 3-4 shift. The progression may not be linear (continuously reduced) or done in proportionate stages for various reasons, including a special need for a gear to reach a specific speed or RPM for passing, racing and so on, or simply economic necessity that the parts were available.

Range and progression are not mutually exclusive, but each limits the number of options for the other. A wide range, which gives a strong torque multiplication in 1st gear for excellent manners in low-speed traffic, especially with a smaller motor, heavy vehicle, or numerically low axle ratio such as 2.50, means that the progression percentages must be high. The amount of engine speed, and therefore power, lost on each up-shift is greater than would be the case in a transmission with less range, but less power in 1st gear. A numerically low 1st gear, such as 2:1, reduces available torque in 1st gear, but allows more choices of progression.

There is no optimal choice of transmission gear ratios or a final drive ratio for best performance at all speeds, as gear ratios are compromises, and not necessarily better than the original ratios for certain purposes.

Brake

Disc brake on a motorcycle

A brake is a mechanical device that inhibits motion by absorbing energy from a moving system. It is used for slowing or stopping a moving vehicle, wheel, axle, or to prevent its motion, most often accomplished by means of friction.

Background

Most brakes commonly use friction between two surfaces pressed together to convert the kinetic energy of the moving object into heat, though other methods of energy conversion may be employed. For example, regenerative braking converts much of the energy to electrical energy, which may be stored for later use. Other methods convert kinetic energy into potential energy in such stored forms as pressurized air or pressurized oil. Eddy current brakes use magnetic fields to convert kinetic energy into electric current in the brake disc, fin, or rail, which is converted into heat. Still other braking methods even transform kinetic energy into different forms, for example by transferring the energy to a rotating flywheel.

Brakes are generally applied to rotating axles or wheels, but may also take other forms such as the surface of a moving fluid (flaps deployed into water or air). Some vehicles use a combination of braking mechanisms, such as drag racing cars with both wheel brakes and a parachute, or airplanes with both wheel brakes and drag flaps raised into the air during landing.

Since kinetic energy increases quadratically with velocity (), an object moving at 10 m/s has 100 times as much energy as one of the same mass moving at 1 m/s, and consequently the theoretical braking distance, when braking at the traction limit, is 100 times as long. In practice, fast vehicles usually have significant air drag, and energy lost to air drag rises quickly with speed.

Almost all wheeled vehicles have a brake of some sort. Even baggage carts and shopping carts may have them for use on a moving ramp. Most fixed-wing aircraft are fitted with wheel brakes on the undercarriage. Some aircraft also feature air brakes designed to reduce their speed in flight. Notable examples include gliders and some World War II-era aircraft, primarily some fighter aircraft and many dive bombers of the era. These allow the aircraft to maintain a safe speed in a steep descent. The Saab B 17 dive bomber and Vought F4U Corsair fighter used the deployed undercarriage as an air brake.

Friction brakes on automobiles store braking heat in the drum brake or disc brake while braking then conduct it to the air gradually. When traveling downhill some vehicles can use their engines to brake.

When the brake pedal of a modern vehicle with hydraulic brakes is pushed against the master cylinder, ultimately a piston pushes the brake pad against the brake disc which slows the wheel down. On the brake drum it is similar as the cylinder pushes the brake shoes against the drum which also slows the wheel down.

Types

Brakes may be broadly described as using friction, pumping, or electromagnetics. One brake may use several principles: for example, a pump may pass fluid through an orifice to create friction:

Rendering of a drum brake

Single pivot side-pull bicycle caliper brake.

Frictional

Frictional brakes are most common and can be divided broadly into "shoe" or "pad" brakes, using an explicit wear surface, and hydrodynamic brakes, such as parachutes, which use friction in a working fluid and do not explicitly wear. Typically the term "friction brake" is used to mean pad/shoe brakes and excludes hydrodynamic brakes, even though hydrodynamic brakes use friction. Friction (pad/shoe) brakes are often rotating devices with a stationary pad and a rotating wear surface. Common configurations include shoes that contract to rub on the outside of a rotating drum, such as a band brake; a rotating drum with shoes that expand to rub the inside of a drum, commonly called a "drum brake", although other drum configurations are possible; and pads that pinch a rotating disc, commonly called a "disc brake". Other brake configurations are used, but less often. For example, PCC trolley brakes include a flat shoe which is clamped to the rail with an electromagnet; the Murphy brake pinches a rotating drum, and the Ausco Lambert disc brake uses a hollow disc (two parallel discs with a structural bridge) with shoes that sit between the disc surfaces and expand laterally.

A drum brake is a vehicle brake in which the friction is caused by a set of brake shoes that press against the inner surface of a rotating drum. The drum is connected to the rotating roadwheel hub.

Drum brakes generally can be found on older car and truck models. However, because of their low production cost, drum brake setups are also installed on the rear of some low-cost newer vehicles. Compared to modern disc brakes, drum brakes wear out faster due to their tendency to overheat.

The disc brake is a device for slowing or stopping the rotation of a road wheel. A brake disc (or rotor in U.S. English), usually made of cast iron or ceramic, is connected to the wheel or the axle. To stop the wheel, friction material in the form of brake pads (mounted in a device called a brake caliper) is forced mechanically, hydraulically, pneumatically or electromagnetically against both sides of the disc. Friction causes the disc and attached wheel to slow or stop.

Ceramic brakes, also called "carbon ceramic", are high-end type of frictional brakes with brake pads and rotors made from porcelain compound blends, that feature better stopping capability and greater resistance to overheat. Due to their high production cost, ceramic brakes aren't widely used as a factory equipment, and their availability on the automotive aftermarket is low compared to traditional metallic brakes. However, being a performance-oriented equipment, ceramic brakes are popular among racers.

Pumping

Pumping brakes are often used where a pump is already part of the machinery. For example, an internal-combustion piston motor can have the fuel supply stopped, and then internal pumping losses of the engine create some braking. Some engines use a valve override called a Jake brake to greatly increase pumping losses. Pumping brakes can dump energy as heat, or can be regenerative brakes that recharge a pressure reservoir called a hydraulic accumulator.

Electromagnetic

Electromagnetic brakes are likewise often used where an electric motor is already part of the machinery. For example, many hybrid gasoline/electric vehicles use the electric motor as a generator to charge electric batteries and also as a regenerative brake. Some diesel/electric railroad locomotives use the electric motors to generate electricity which is then sent to a resistor bank and dumped as heat. Some vehicles, such as some transit buses, do not already have an electric motor but use a secondary "retarder" brake that is effectively a generator with an internal short-circuit. Related types of such a brake are eddy current brakes, and electro-mechanical brakes (which actually are magnetically driven friction brakes, but nowadays are often just called "electromagnetic brakes" as well).

Electromagnetic brakes slow an object through electromagnetic induction, which creates resistance and in turn either heat or electricity. Friction brakes apply pressure on two separate objects to slow the vehicle in a controlled manner.

Characteristics

Brakes are often described according to several characteristics including:

- Peak force – The peak force is the maximum decelerating effect that can be obtained. The peak force is often greater than the traction limit of the tires, in which case the brake can cause a wheel skid.

- Continuous power dissipation – Brakes typically get hot in use, and fail when the temperature gets too high. The greatest amount of power (energy per unit time) that can be dissipated through the brake without failure is the continuous power dissipation. Continuous power dissipation often depends on e.g., the temperature and speed of ambient cooling air.

- Fade – As a brake heats, it may become less effective, called brake fade. Some designs are inherently prone to fade, while other designs are relatively immune. Further, use considerations, such as cooling, often have a big effect on fade.

- Smoothness – A brake that is grabby, pulses, has chatter, or otherwise exerts varying brake force may lead to skids. For example, railroad wheels have little traction, and friction brakes without an anti-skid mechanism often lead to skids, which increases maintenance costs and leads to a "thump thump" feeling for riders inside.

- Power – Brakes are often described as "powerful" when a small human application force leads to a braking force that is higher than typical for other brakes in the same class. This notion of "powerful" does not relate to continuous power dissipation, and may be confusing in that a brake may be "powerful" and brake strongly with a gentle brake application, yet have lower (worse) peak force than a less "powerful" brake.

- Pedal feel – Brake pedal feel encompasses subjective perception of brake power output as a function of pedal travel. Pedal travel is influenced by the fluid displacement of the brake and other factors.

- Drag – Brakes have varied amount of drag in the off-brake condition depending on design of the system to accommodate total system compliance and deformation that exists under braking with ability to retract friction material from the rubbing surface in the off-brake condition.

- Durability – Friction brakes have wear surfaces that must be renewed periodically. Wear surfaces include the brake shoes or pads, and also the brake disc or drum. There may be tradeoffs, for example a wear surface that generates high peak force may also wear quickly.

- Weight – Brakes are often "added weight" in that they serve no other function. Further, brakes are often mounted on wheels, and unsprung weight can significantly hurt traction in some circumstances. "Weight" may mean the brake itself, or may include additional support structure.

- Noise – Brakes usually create some minor noise when applied, but often create squeal or grinding noises that are quite loud.

Brake Boost

Brake booster from a Geo Storm.

Most modern vehicles use a vacuum assisted brake system that greatly increases the force applied to the vehicle's brakes by its operator. This additional force is supplied by the manifold vacuum generated by air flow being obstructed by the throttle on a running engine. This force is greatly reduced when the engine is running at fully open throttle, as the difference between ambient air pressure and manifold (absolute) air pressure is reduced, and therefore available vacuum is diminished. However, brakes are rarely applied at full throttle; the driver takes the right foot off the gas pedal and moves it to the brake pedal - unless left-foot braking is used.

Because of low vacuum at high RPM, reports of unintended acceleration are often accompanied by complaints of failed or weakened brakes, as the high-revving engine, having an open throttle, is unable to provide enough vacuum to power the brake booster. This problem is exacerbated in vehicles equipped with automatic transmissions as the vehicle will automatically downshift upon application of the brakes, thereby increasing the torque delivered to the driven-wheels in contact with the road surface.

Noise

Although ideally a brake would convert all the kinetic energy into heat, in practice a significant amount may be converted into acoustic energy instead, contributing to noise pollution.

For road vehicles, the noise produced varies significantly with tire construction, road surface, and the magnitude of the deceleration. Noise can be caused by different things. These are signs that there may be issues with brakes wearing out over time.

Brake lever on a horse-drawn hearse

Fires

Railway braking produces sparks and is an important cause of forest fires.

Inefficiency

A significant amount of energy is always lost while braking, even with regenerative braking which is not perfectly efficient. Therefore, a good metric of efficient energy use while driving is to note how much one is braking. If the majority of deceleration is from unavoidable friction instead of braking, one is squeezing out most of the service from the vehicle. Minimizing brake use is one of the fuel economy-maximizing behaviors.

While energy is always lost during a brake event, a secondary factor that influences efficiency is "off-brake drag", or drag that occurs when the brake is not intentionally actuated. After a braking event, hydraulic pressure drops in the system, allowing the brake caliper pistons to retract. However, this retraction must accommodate all compliance in the system (under pressure) as well as thermal distortion of components like the brake disc or the brake system will drag until the contact with the disc, for example, knocks the pads and pistons back from the rubbing surface. During this time, there can be significant brake drag. This brake drag can lead to significant parasitic power loss, thus impact fuel economy and overall vehicle performance.

Clutch

Single, dry clutch friction disc. The splined hub is attached to the disc with springs to damp chatter.

A clutch is a mechanical device that engages and disengages the power transmission, especially from driving shaft to driven shaft.

Clutches are used whenever the transmission of power or motion must be controlled either in amount or over time (e.g., electric screwdrivers limit how much torque is transmitted through use of a clutch; clutches control whether automobiles transmit engine power to the wheels).

In the simplest application, clutches connect and disconnect two rotating shafts (drive shafts or line shafts). In these devices, one shaft is typically attached to an engine or other power unit (the driving member) while the other shaft (the driven member) provides output power for work. While typically the motions involved are rotary, linear clutches are also possible.

In a torque-controlled drill, for instance, one shaft is driven by a motor and the other drives a drill chuck. The clutch connects the two shafts so they may be locked together and spin at the same speed (engaged), locked together but spinning at different speeds (slipping), or unlocked and spinning at different speeds (disengaged).

Inter-locking Parts clutches

This type of clutch has protruding circular edge and a hole for them that engages and disengages during operation, less shaft force and greater circular force. This type is less effective since human foot or hand power on clutching reaches about 10 KN or 1,000,000 kGs.

Friction Clutches

The vast majority of clutches ultimately rely on frictional forces for their operation. The purpose of friction clutches is to connect a moving member to another that is moving at a different speed or stationary, often to synchronize the speeds, and/or to transmit power. Usually, as little slippage (difference in speeds) as possible between the two members is desired.

A friction clutch

Materials

Various materials have been used for the disc-friction facings, including asbestos in the past. Modern clutches typically use a compound organic resin with copper wire facing or a ceramic material. Ceramic materials are typically used in heavy applications such as racing or heavy-duty hauling, though the harder ceramic materials increase flywheel and pressure plate wear.

In the case of "wet" clutches, composite paper materials are very common. Since these "wet" clutches typically use an oil bath or flow-through cooling method for keeping the disc pack lubricated and cooled, very little wear is seen when using composite paper materials.

Push/Pull

Friction-disc clutches generally are classified as *push type* or *pull type* depending on the location of the pressure plate fulcrum points. In a pull-type clutch, the action of pressing the pedal pulls the release bearing, pulling on the diaphragm spring and disengaging the vehicle drive. The opposite is true with a push type, the release bearing is pushed into the clutch disengaging the vehicle drive. In this instance, the release bearing can be known as a thrust bearing (as per the image above).

Dampers

A clutch damper is a device that softens the response of the clutch engagement/disengagement. In automotive applications, this is often provided by a mechanism in the clutch disc centres. In addition to the damped disc centres, which reduce driveline vibration, pre-dampers may be used to reduce gear rattle at idle by changing the natural frequency of the disc. These weaker springs are compressed solely by the radial vibrations of an idling engine. They are fully compressed and no longer in use once the main damper springs take up drive.

Load

Mercedes truck examples: A clamp load of 33 kN is normal for a single plate 430. The 400 Twin application offers a clamp load of a mere 23 kN. Bursts speeds are typically around 5,000 rpm with the weakest point being the facing rivet.

Manufacturing

Modern clutch development focuses its attention on the simplification of the overall assembly and/or manufacturing method. For example, drive straps are now commonly employed to transfer torque as well as lift the pressure plate upon disengagement of vehicle drive. With regard to the manufacture of diaphragm springs, heat treatment is crucial. Laser welding is becoming more common as a method of attaching the drive plate to the disc ring with the laser typically being between 2-3KW and a feed rate 1m/minute.

Multiple Plate Clutch

This type of clutch has several driving members interleaved or "stacked" with several driven members. It is used in racing cars including Formula 1, IndyCar, World Rally and even most club racing. Multiplate clutches see much use in drag racing, which requires the best acceleration possible, and is notorious for the abuse the clutch is subjected to. Thus motorcycles, automatic transmissions and in some diesel locomotives with mechanical transmissions. It is also used in some electronically controlled all-wheel drive systems as well as in some transfer cases. They can also be found in some heavy machinery such as tanks and AFV's (T-54) and earthmoving equipment (front-end loaders, bulldozers), as well as components in certain types of limited slip differentials. The benefit in the case of motorsports is that you can achieve the same total friction force with a much smaller overall diameter (or conversely, a much greater friction force for the same diameter, important in cases where a vehicle is modified with greater power, yet the maximum physical size of the clutch unit is constrained by the clutch housing). In motorsports vehicles that run at high engine/drivetrain speeds, the smaller diameter reduces rotational inertia, making the drivetrain components accelerate more rapidly, as well as reducing the angular velocity of the outer areas of the clutch unit, which could become highly stressed and fail at the extremely high drivetrain rotational rates achieved in sports such as Formula 1 or drag racing. In the case of heavy equipment, which often deal with very high torque forces and drivetrain loads, a single plate clutch of the necessary strength would be too large to easily package as a component of the driveline.

Another, different theme on the multiplate clutch is the clutches used in the fastest classes of drag racing, highly specialized, purpose-built cars such as Top Fuel or Funny Cars. These cars are so powerful that to attempt a start with a simple clutch would result in complete loss of traction. To avoid this problem, Top Fuel cars actually use a single, fixed gear ratio, and a *series* of clutches that are engaged one at a time, rather than in unison, progressively allowing more power to the wheels. A single one of these clutch plates (as designed) can not hold more than a fraction of the power of the engine, so the driver starts with only the first clutch engaged. This clutch is overwhelmed by the power of the engine, allowing only a fraction of the power to the wheels, much like "slipping the clutch" in a slower car, but working not requiring concentration from the driver. As speed builds, the driver pulls a lever, which engages a second clutch, sending a bit more of the engine power to the wheels, and so on. This continues through several clutches until the car has reached a speed where the last clutch can be engaged. With all clutches engaged, the engine is now sending all of its power to the rear wheels. This is far more predictable and repeatable than the driver manually slipping the clutch himself and then shifting through the gears, given the extreme violence of the run and the speed at which is all unfolds. Another benefit is that there is no need to break the power flow in order to swap gears (a conventional manual cannot transmit power while between gears, which is important because 1/100ths of a second are important in Top Fuel races). A traditional multiplate clutch would be more prone to overheating and failure, as all the plates must be subjected to heat and friction together until the clutch is fully engaged, while a Top Fuel car keeps its last clutches in "reserve" until the cars speed allows full engagement. It is relatively easy to design the last stages to be much more powerful than the first, in order to ensure they can absorb the power of the engine even if the first clutches burn out or overheat from the extreme friction.

Wet Vs. Dry Systems

A *wet clutch* is immersed in a cooling lubricating fluid that also keeps surfaces clean and provides smoother performance and longer life. Wet clutches, however, tend to lose some energy to the liquid. Since the surfaces of a wet clutch can be slippery (as with a motorcycle clutch bathed in engine oil), stacking multiple clutch discs can compensate for the lower coefficient of friction and so eliminate slippage under power when fully engaged. The Hele-Shaw clutch was a wet clutch that relied entirely on viscous effects, rather than on friction.

A *dry clutch*, as the name implies, is not bathed in liquid and uses friction to engage.

Centrifugal Clutch

A centrifugal clutch is used in some vehicles (e.g., mopeds) and also in other applications where the speed of the engine defines the state of the clutch, for example, in a chainsaw. This clutch system employs centrifugal force to automatically engage the clutch when the engine rpm rises above a threshold and to automatically disengage the clutch when the engine rpm falls low enough. The system involves a clutch shoe or shoes attached to the driven shaft, rotating inside a clutch bell attached to the output shaft. The shoe(s) are held inwards by springs until centrifugal force overcomes the spring tension and the shoe(s) make contact with the bell, driving the output. In the case of a chainsaw this allows the chain to remain stationary whilst the engine is idling; once the throttle is pressed and the engine speed rises, the centrifugal clutch engages and the cutting chain moves.

Cone Clutch

As the name implies, a cone clutch has conical friction surfaces. The cone's taper means that a given amount of movement of the actuator makes the surfaces approach (or recede) much more slowly than in a disc clutch. As well, a given amount of actuating force creates more pressure on the mating surfaces. The best known example of a cone clutch is a synchronizer ring in a manual transmission. The synchronizer ring is responsible for "synchronizing" the speeds of the shift hub and the gear wheel to ensure a smooth gear change.

Torque Limiter

Also known as a slip clutch or *safety clutch*, this device allows a rotating shaft to slip when higher than normal resistance is encountered on a machine. An example of a safety clutch is the one mounted on the driving shaft of a large grass mower. The clutch yields if the blades hit a rock, stump, or other immobile object, thus avoiding a potentially damaging torque transfer to the engine, possibly twisting or fracturing the crankshaft.

Motor-driven mechanical calculators had these between the drive motor and gear train, to limit damage when the mechanism jammed, as motors used in such calculators had high stall torque and were capable of causing damage to the mechanism if torque wasn't limited.

Carefully designed clutches operate, but continue to transmit maximum permitted torque, in such tools as controlled-torque screwdrivers.

Non-Slip Clutches

Some clutches are not designed to slip; torque may only be transmitted either fully engaged or disengaged to avoid catastrophic damage. An example of this is the dog clutch, most commonly used in non-synchromesh transmissions.

Major Types by Application

Vehicular (General)

There are different designs of vehicle clutch but most are based on one or more friction discs pressed tightly together or against a flywheel using springs. The friction material varies in composition depending on many considerations such as whether the clutch is "dry" or "wet". Friction discs once contained asbestos but this has been largely eliminated. Clutches found in heavy duty applications such as trucks and competition cars use ceramic plates that have a greatly increased friction coefficient. However, these have a "grabby" action generally considered unsuitable for passenger cars. The spring pressure is released when the clutch pedal is depressed thus either pushing or pulling the diaphragm of the pressure plate, depending on type. However, raising the engine speed too high while engaging the clutch causes excessive clutch plate wear. Engaging the clutch abruptly when the engine is turning at high speed causes a harsh, jerky start. This kind of start is necessary and desirable in drag racing and other competitions, where speed is more important than comfort.

Automobile Powertrain

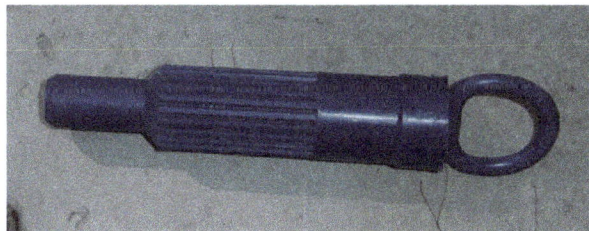

This plastic pilot shaft guide tool is used to align the clutch disk as the spring-loaded pressure plate is installed. The transmission's drive splines and pilot shaft have a complementary shape. A number of such devices fit various makes and models of drivetrains.

In a modern car with a manual transmission the clutch is operated by the left-most pedal using a hydraulic or cable connection from the pedal to the clutch mechanism. On older cars the clutch might be operated by a mechanical linkage. Even though the clutch may physically be located very close to the pedal, such remote means of actuation are necessary to eliminate the effect of vibrations and slight engine movement, engine mountings being flexible by design. With a rigid mechanical linkage, smooth engagement would be near-impossible because engine movement inevitably occurs as the drive is "taken up."

The default state of the clutch is *engaged* - that is the connection between engine and gearbox is always "on" unless the driver presses the pedal and disengages it. If the engine is running with clutch engaged and the transmission in neutral, the engine spins the input shaft of the transmission, but no power is transmitted to the wheels.

The clutch is located between the engine and the gearbox, as disengaging it is required to change gear. Although the gearbox does not stop rotating during a gear change, there is no torque transmitted through it, thus less friction between gears and their engagement dogs. The output shaft of the gearbox is permanently connected to the final drive, then the wheels, and so both always rotate together, at a fixed speed ratio. With the clutch disengaged, the gearbox input shaft is free to change its speed as the internal ratio is changed. Any resulting difference in speed between the engine and gearbox is evened out as the clutch slips slightly during re-engagement.

Clutches in typical cars are mounted directly to the face of the engine's flywheel, as this already provides a convenient large diameter steel disk that can act as one driving plate of the clutch. Some racing clutches use small multi-plate disk packs that are not part of the flywheel. Both clutch and flywheel are enclosed in a conical bellhousing, which (in a rear-wheel drive car) usually forms the main mounting for the gearbox.

A few cars, notably the Alfa Romeo Alfetta, Porsche 924, and Chevrolet Corvette (since 1997), sought a more even weight distribution between front and back by placing the weight of the transmission at the rear of the car, combined with the rear axle to form a transaxle. The clutch was mounted with the transaxle and so the propeller shaft rotated continuously with the engine, even when in neutral gear or declutched.

Motorcycles

A basket clutch

Motorcycles typically employ a wet clutch with the clutch riding in the same oil as the transmission. These clutches are usually made up of a stack of alternating plain steel and friction plates. Some plates have lugs on their inner diameters that lock them to the engine crankshaft. Other plates have lugs on their outer diameters that lock them to a basket that turns the transmission input shaft. A set of coil springs or a diaphragm spring plate force the plates together when the clutch is engaged.

On motorcycles the clutch is operated by a hand lever on the left handlebar. No pressure on the lever means that the clutch plates are engaged (driving), while pulling the lever back towards the rider disengages the clutch plates through cable or hydraulic actuation, allowing the rider to shift gears or coast. Racing motorcycles often use slipper clutches to eliminate the effects of engine braking, which, being applied only to the rear wheel, can cause instability.

Automobile Non-Powertrain

Cars use clutches in places other than the drive train. For example, a belt-driven engine cooling fan may have a heat-activated clutch. The driving and driven members are separated by a silicone-based fluid and a valve controlled by a bimetallic spring. When the temperature is low, the spring winds and closes the valve, which lets the fan spin at about 20% to 30% of the shaft speed. As the temperature of the spring rises, it unwinds and opens the valve, allowing fluid past the valve, makes the fan spin at about 60% to 90% of shaft speed. Other clutches—such as for an air conditioning compressor—electronically engage clutches using magnetic force to couple the driving member to the driven member.

Other Clutches and Applications

- Belt clutch: Used on agricultural equipment, lawn mowers, tillers, and snow blowers. Engine power is transmitted via a set of belts that are slack when the engine is idling, but an idler pulley can tighten the belts to increase friction between the belts and the pulleys.

- Dog clutch: Utilized in automobile manual transmissions mentioned above. Positive engagement, non-slip. Typically used where slipping is not acceptable and space is limited. Partial engagement under any significant load can be destructive.

- Hydraulic clutch: The driving and driven members are not in physical contact; coupling is hydrodynamic.

- Electromagnetic clutch are, typically, engaged by an electromagnet that is an integral part of the clutch assembly. Another type, *magnetic particle clutches*, contain magnetically influenced particles in a chamber between driving and driven members—application of direct current makes the particles clump together and adhere to the operating surfaces. Engagement and slippage are notably smooth.

- Overrunning clutch or freewheel: If some external force makes the driven member rotate faster than the driver, the clutch effectively disengages. Examples include:

    - Borg-Warner overdrive transmissions in cars

    - Ratchet: typical bicycles have these so that the rider can stop pedaling and coast

    - An oscillating member where this clutch can then convert the oscillations into intermittent linear or rotational motion of the complimentary member; others use ratchets with the pawl mounted on a moving member

    - The winding knob of a camera employs a (silent) wrap-spring type as a clutch in winding and as a brake in preventing it from being turned backwards.

- The rotor drive train in helicopters uses a freewheeling clutch to disengage the rotors from the engine in the event of engine failure, allowing the craft to safely descend by autorotation.

- Wrap-spring clutches: These have a helical spring, typically wound with square-cross-section wire. These were developed in the late 19th and early 20th century. In simple form the spring is fastened at one end to the driven member; its other end is unattached. The spring fits closely around a cylindrical driving member. If the driving member rotates in the direction that would unwind the spring the spring expands minutely and slips although with some drag. Because of this, spring clutches must typically be lubricated with light oil. Rotating the driving member the other way makes the spring wrap itself tightly around the driving surface and the clutch locks up very quickly. The torque required to make a spring clutch slip grows exponentially with the number of turns in the spring, obeying the capstan equation.

Specialty Clutches and Applications

Single-revolution Clutch

Single-revolution clutches were developed in the 19th century to power machinery such as shears or presses where a single pull of the operating lever or (later) press of a button would trip the mechanism, engaging the clutch between the power source and the machine's crankshaft for exactly one revolution before disengaging the clutch. When the clutch is disengaged and the driven member is stationary. Early designs were typically dog clutches with a cam on the driven member used to disengage the dogs at the appropriate point.

Greatly simplified single-revolution clutches were developed in the 20th century, requiring much smaller operating forces and in some variations, allowing for a fixed fraction of a revolution per operation. Fast action friction clutches replaced dog clutches in some applications, eliminating the problem of impact loading on the dogs every time the clutch engaged.

In addition to their use in heavy manufacturing equipment, single-revolution clutches were applied to numerous small machines. In tabulating machines, for example, pressing the operate key would trip a single revolution clutch to process the most recently entered number. In typesetting machines, pressing any key selected a particular character and also engaged a single rotation clutch to cycle the mechanism to typeset that character. Similarly, in teleprinters, the receipt of each character tripped a single-revolution clutch to operate one cycle of the print mechanism.

In 1928, Frederick G. Creed developed a single-turn spring clutch that was particularly well suited to the repetitive start-stop action required in teleprinters. In 1942, two employees of Pitney Bowes Postage Meter Company developed an improved single turn spring clutch. In these clutches, a coil spring is wrapped around the driven shaft and held in an expanded configuration by the trip lever. When tripped, the spring rapidly contracts around the power shaft engaging the clutch. At the end of one revolution, if the trip lever has been reset, it catches the end of the spring (or a pawl attached to it) and the angular momentum of the driven member releases the tension on the spring. These clutches have long operating lives, many have cycled for tens and perhaps hundreds of millions of cycles without need of maintenance other than occasional lubrication.

Cascaded-pawl Single-revolution Clutches

Cascaded-pawl single-revolution clutch driving the cam cluster in a Teletype Model 33 that performs fully mechanical conversion of incoming asynchronous serial data to parallel form. The clutch drum, lower left, has been removed to expose the pawls and trip projection.

These superseded wrap-spring single-revolution clutches in page printers, such as teleprinters, including the Teletype Model 28 and its successors, using the same design principles. IBM Selectric typewriters also used them. These are typically disc-shaped assemblies mounted on the driven shaft. Inside the hollow disc-shaped drive drum are two or three freely floating pawls arranged so that when the clutch is tripped, the pawls spring outward much like the shoes in a drum brake. When engaged, the load torque on each pawl transfers to the others to keep them engaged. These clutches do not slip once locked up, and they engage very quickly, on the order of milliseconds. A trip projection extends out from the assembly. If the trip lever engaged this projection, the clutch was disengaged. When the trip lever releases this projection, internal springs and friction engage the clutch. The clutch then rotates one or more turns, stopping when the trip lever again engages the trip projection.

Kickback Clutch-brakes

These mechanisms were found in some types of synchronous-motor-driven electric clocks. Many different types of synchronous clock motors were used, including the pre-World War II Hammond manual-start clocks. Some types of self-starting synchronous motors always started when power was applied, but in detail, their behaviour was chaotic and they were equally likely to start rotating in the wrong direction. Coupled to the rotor by one (or possibly two) stages of reduction gearing was a wrap-spring clutch-brake. The spring did not rotate. One end was fixed; the other was free. It rode freely but closely on the rotating member, part of the clock's gear train. The clutch-brake locked up when rotated backwards, but also had some spring action. The inertia of the rotor going backwards engaged the clutch and wound the spring. As it unwound, it restarted the motor in the correct direction. Some designs had no explicit spring as such—but were simply compliant mechanisms. The mechanism was lubricated and wear did not present a problem.

Lock-up Clutch

A Lock-up clutch is used in some automatic transmissions for motor vehicles. Above a certain speed (usually 60 km/h) it locks the torque converter to minimise power loss and improve fuel efficiency.

Bearing (Mechanical)

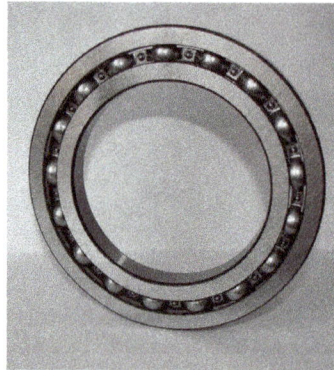

Ball bearing

A bearing is a machine element that constrains relative motion to only the desired motion, and reduces friction between moving parts. The design of the bearing may, for example, provide for free linear movement of the moving part or for free rotation around a fixed axis; or, it may *prevent* a motion by controlling the vectors of normal forces that bear on the moving parts. Many bearings also *facilitate* the desired motion as much as possible, such as by minimizing friction. Bearings are classified broadly according to the type of operation, the motions allowed, or to the directions of the loads (forces) applied to the parts.

The term "bearing" is derived from the verb "to bear"; a bearing being a machine element that allows one part to bear (i.e., to support) another. The simplest bearings are bearing surfaces, cut or formed into a part, with varying degrees of control over the form, size, roughness and location of the surface. Other bearings are separate devices installed into a machine or machine part. The most sophisticated bearings for the most demanding applications are very precise devices; their manufacture requires some of the highest standards of current technology.

History

Tapered roller bearing

Drawing of Leonardo da Vinci (*1452-1519*) Study of a ball bearing

The invention of the rolling bearing, in the form of wooden rollers supporting, or bearing, an object being moved is of great antiquity, and may predate the invention of the wheel.

Though it is often claimed that the Egyptians used roller bearings in the form of tree trunks under sleds, this is modern speculation. They are depicted in their own drawings in the tomb of Djehutihotep as moving massive stone blocks on sledges with liquid-lubricated runners which would constitute a plain bearing. There are also Egyptian drawings of bearings used with hand drills.

The earliest recovered example of a rolling element bearing is a wooden ball bearing supporting a rotating table from the remains of the Roman Nemi ships in Lake Nemi, Italy. The wrecks were dated to 40 BC.

Leonardo da Vinci incorporated drawings of ball bearings in his design for a helicopter around the year 1500. This is the first recorded use of bearings in an aerospace design. However, Agostino Ramelli is the first to have published sketches of roller and thrust bearings. An issue with ball and roller bearings is that the balls or rollers rub against each other causing additional friction which can be reduced by enclosing the balls or rollers within a cage. The captured, or caged, ball bearing was originally described by Galileo in the 17th century.

The first practical caged-roller bearing was invented in the mid-1740s by horologist John Harrison for his H3 marine timekeeper. This uses the bearing for a very limited oscillating motion but Harrison also used a similar bearing in a truly rotary application in a contemporaneous regulator clock.

Industrial Era

The first modern recorded patent on ball bearings was awarded to Philip Vaughan, a British inventor and ironmaster who created the first design for a ball bearing in Carmarthen in 1794. His was the first modern ball-bearing design, with the ball running along a groove in the axle assembly.

Bearings played a pivotal role in the nascent Industrial Revolution, allowing the new industrial machinery to operate efficiently. For example, they saw use for holding wheel and axle to greatly reduce friction over that of dragging an object by making the friction act over a shorter distance as the wheel turned.

The first plain and rolling-element bearings were wood closely followed by bronze. Over their history bearings have been made of many materials including ceramic, sapphire, glass, steel, bronze, other metals and plastic (e.g., nylon, polyoxymethylene, polytetrafluoroethylene, and UHMWPE) which are all used today.

Watch makers produce "jeweled" watches using sapphire plain bearings to reduce friction thus allowing more precise time keeping.

Even basic materials can have good durability. As examples, wooden bearings can still be seen today in old clocks or in water mills where the water provides cooling and lubrication.

Early Timken tapered roller bearing with notched rollers

The first patent for a radial style ball bearing was awarded to Jules Suriray, a Parisian bicycle mechanic, on 3 August 1869. The bearings were then fitted to the winning bicycle ridden by James Moore in the world's first bicycle road race, Paris-Rouen, in November 1869.

In 1883, Friedrich Fischer, founder of FAG, developed an approach for milling and grinding balls of equal size and exact roundness by means of a suitable production machine and formed the foundation for creation of an independent bearing industry.

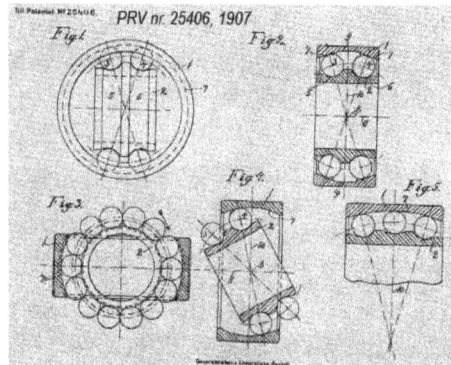

Wingquist original patent of self-aligning ball bearing

The modern, self-aligning design of ball bearing is attributed to Sven Wingquist of the SKF ball-bearing manufacturer in 1907, when he was awarded Swedish patent No. 25406 on its design.

Henry Timken, a 19th-century visionary and innovator in carriage manufacturing, patented the tapered roller bearing in 1898. The following year he formed a company to produce his innovation. Over a century the company grew to make bearings of all types, including specialty steel and an array of related products and services.

Erich Franke invented and patented the wire race bearing in 1934. His focus was on a bearing design with a cross section as small as possible and which could be integrated into the enclosing design. After World War II he founded together with Gerhard Heydrich the company Franke & Heydrich KG (today Franke GmbH) to push the development and production of wire race bearings.

Richard Stribeck's extensive research on ball bearing steels identified the metallurgy of the commonly used 100Cr6 (AISI 52100) showing coefficient of friction as a function of pressure.

Designed in 1968 and later patented in 1972, Bishop-Wisecarver's co-founder Bud Wisecarver created vee groove bearing guide wheels, a type of linear motion bearing consisting of both an external and internal 90-degree vee angle.

In the early 1980s, Pacific Bearing's founder, Robert Schroeder, invented the first bi-material plain bearing which was size interchangeable with linear ball bearings. This bearing had a metal shell (aluminum, steel or stainless steel) and a layer of Teflon-based material connected by a thin adhesive layer.

Today ball and roller bearings are used in many applications which include a rotating component. Examples include ultra high speed bearings in dental drills, aerospace bearings in the Mars Rover, gearbox and wheel bearings on automobiles, flexure bearings in optical alignment systems, bicycle wheel hubs, and air bearings used in Coordinate-measuring machines.

Common

By far, the most common bearing is the plain bearing, a bearing which uses surfaces in rubbing contact, often with a lubricant such as oil or graphite. A plain bearing may or may not be a discrete device. It may be nothing more than the bearing surface of a hole with a shaft passing through it, or of a planar surface that bears another (in these cases, not a discrete device); or it may be a layer of bearing metal either fused to the substrate (semi-discrete) or in the form of a separable sleeve (discrete). With suitable lubrication, plain bearings often give entirely acceptable accuracy, life, and friction at minimal cost. Therefore, they are very widely used.

However, there are many applications where a more suitable bearing can improve efficiency, accuracy, service intervals, reliability, speed of operation, size, weight, and costs of purchasing and operating machinery.

Thus, there are many types of bearings, with varying shape, material, lubrication, principle of operation, and so on.

Types

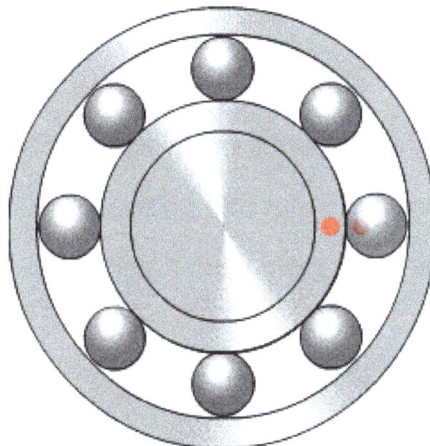

Animation of ball bearing (without a cage). The inner ring rotates and the outer ring is stationary.

There are at least 6 common types of bearing, each of which operates on different principles:

- Plain bearing, also known by the specific styles: bushing, journal bearing, sleeve bearing, rifle bearing, composite bearing.

- Rolling-element bearing such as ball bearings and roller bearings

- Jewel bearing, in which the load is carried by rolling the axle slightly off-center

- Fluid bearing, in which the load is carried by a gas or liquid

- Magnetic bearing, in which the load is carried by a magnetic field

- Flexure bearing, in which the motion is supported by a load element which bends.

Motions

- Common motions permitted by bearings are:
- axial rotation e.g. shaft rotation
- linear motion e.g. drawer
- spherical rotation e.g. ball and socket joint
- hinge motion e.g. door, elbow, knee

Friction

Reducing friction in bearings is often important for efficiency, to reduce wear and to facilitate extended use at high speeds and to avoid overheating and premature failure of the bearing. Essentially, a bearing can reduce friction by virtue of its shape, by its material, or by introducing and containing a fluid between surfaces or by separating the surfaces with an electromagnetic field.

By shape, gains advantage usually by using spheres or rollers, or by forming flexure bearings.

By material, exploits the nature of the bearing material used. (An example would be using plastics that have low surface friction.)

By fluid, exploits the low viscosity of a layer of fluid, such as a lubricant or as a pressurized medium to keep the two solid parts from touching, or by reducing the normal force between them.

By fields, exploits electromagnetic fields, such as magnetic fields, to keep solid parts from touching.

""Air pressure"" exploits air pressure to keep solid parts from touching.

Combinations of these can even be employed within the same bearing. An example of this is where the cage is made of plastic, and it separates the rollers/balls, which reduce friction by their shape and finish.

Loads

Bearings vary greatly over the size and directions of forces that they can support.

Forces can be predominately radial, axial (thrust bearings) or bending moments perpendicular to the main axis.

Speeds

Different bearing types have different operating speed limits. Speed is typically specified as maximum relative surface speeds, often specified ft/s or m/s. Rotational bearings typically describe performance in terms of the product DN where D is the mean diameter (often in mm) of the bearing and N is the rotation rate in revolutions per minute.

Generally there is considerable speed range overlap between bearing types. Plain bearings typically handle only lower speeds, rolling element bearings are faster, followed by fluid bearings and finally magnetic bearings which are limited ultimately by centripetal force overcoming material strength.

Play

Some applications apply bearing loads from varying directions and accept only limited play or "slop" as the applied load changes. One source of motion is gaps or "play" in the bearing. For example, a 10 mm shaft in a 12 mm hole has 2 mm play.

Allowable play varies greatly depending on the use. As example, a wheelbarrow wheel supports radial and axial loads. Axial loads may be hundreds of newtons force left or right, and it is typically acceptable for the wheel to wobble by as much as 10 mm under the varying load. In contrast, a lathe may position a cutting tool to ±0.02 mm using a ball lead screw held by rotating bearings. The bearings support axial loads of thousands of newtons in either direction, and must hold the ball lead screw to ±0.002 mm across that range of loads

Stiffness

A second source of motion is elasticity in the bearing itself. For example, the balls in a ball bearing are like stiff rubber, and under load deform from round to a slightly flattened shape. The race is also elastic and develops a slight dent where the ball presses on it.

The stiffness of a bearing is how the distance between the parts which are separated by the bearing varies with applied load. With rolling element bearings this is due to the strain of the ball and race. With fluid bearings it is due to how the pressure of the fluid varies with the gap (when correctly loaded, fluid bearings are typically stiffer than rolling element bearings).

Service Life

Fluid and Magnetic Bearings

Fluid and magnetic bearings can have practically indefinite service lives. In practice, there are fluid bearings supporting high loads in hydroelectric plants that have been in nearly continuous service since about 1900 and which show no signs of wear.

Rolling Element Bearings

Rolling element bearing life is determined by load, temperature, maintenance, lubrication, material defects, contamination, handling, installation and other factors. These factors can all have a significant effect on bearing life. For example, the service life of bearings in one application was extended dramatically by changing how the bearings were stored before installation and use, as vibrations during storage caused lubricant failure even when the only load on the bearing was its own weight; the resulting damage is often false brinelling. Bearing life is statistical: several samples of a given bearing will often exhibit a bell curve of service life, with a few samples showing significantly better or worse life. Bearing life varies because microscopic structure and contamination vary greatly even where macroscopically they seem identical.

L10 Life

Bearings are often specified to give an "L10" life (outside the USA, it may be referred to as "B10" life.) This is the life at which ten percent of the bearings in that application can be expected to have failed due to classical fatigue failure (and not any other mode of failure like lubrication starvation, wrong mounting etc.), or, alternatively, the life at which ninety percent will still be operating.The L10 life of the bearing is theoretical life and may not represent service life of the bearing. Bearings are also rated using C_o (static loading) value. This is the basic load rating as a reference, and not an actual load value.

Plain Bearings

For plain bearings some materials give much longer life than others. Some of the John Harrison clocks still operate after hundreds of years because of the *lignum vitae* wood employed in their construction, whereas his metal clocks are seldom run due to potential wear.

Flexure Bearings

Flexure bearings rely on elastic properties of material.Flexure bearings bend a piece of material repeatedly. Some materials fail after repeated bending, even at low loads, but careful material selection and bearing design can make flexure bearing life indefinite.

Short-Life Bearings

Although long bearing life is often desirable, it is sometimes not necessary. Tedric A. Harris describes a bearing for a rocket motor oxygen pump that gave several hours life, far in excess of the several tens of minutes life needed.

Composite Bearings

Depending on the customized specifications (backing material and PTFE compounds), composite bearings can operate up to 30 years without maintenance.

External Factors

The service life of the bearing is affected by many parameters that are not controlled by the bearing manufacturers. For example, bearing mounting, temperature, exposure to external environment, lubricant cleanliness and electrical currents through bearings etc. The disruption from PWM inverter which are generating high frequency motorbearing currents can be suppressed by inductive absorbers like CoolBLUE cores, which need to be put over the three phases giving a high frequency impedance against the common mode or motorbearing currents.

The temperature and terrain of the micro-surface will determine the amount of friction by the touching of solid parts.

Certain elements and fields reduce friction, while increasing speeds.

Strength and mobility help determine the amount of load the bearing type can carry.

Alignment factors can play a damaging role in wear and tear, yet overcome by computer aid signaling and non-rubbing bearing types, such as magnetic levitation or air field pressure.

Maintenance and Lubrication

Many bearings require periodic maintenance to prevent premature failure, but many others require little maintenance. The latter include various kinds of fluid and magnetic bearings, as well as rolling-element bearings that are described with terms including *sealed bearing* and *sealed for life*. These contain seals to keep the dirt out and the grease in. They work successfully in many applications, providing maintenance-free operation. Some applications cannot use them effectively.

Nonsealed bearings often have a grease fitting, for periodic lubrication with a grease gun, or an oil cup for periodic filling with oil. Before the 1970s, sealed bearings were not encountered on most machinery, and oiling and greasing were a more common activity than they are today. For example, automotive chassis used to require "lube jobs" nearly as often as engine oil changes, but today's car chassis are mostly sealed for life. From the late 1700s through mid 1900s, industry relied on many workers called oilers to lubricate machinery frequently with oil cans.

Factory machines today usually have *lube systems*, in which a central pump serves periodic charges of oil or grease from a reservoir through *lube lines* to the various *lube points* in the machine's bearing surfaces, bearing journals, pillow blocks, and so on. The timing and number of such *lube cycles* is controlled by the machine's computerized control, such as PLC or CNC, as well as by manual override functions when occasionally needed. This automated process is how all modern CNC machine tools and many other modern factory machines are lubricated. Similar lube systems are also used on nonautomated machines, in which case there is a hand pump that a machine operator is supposed to pump once daily (for machines in constant use) or once weekly. These are called *one-shot systems* from their chief selling point: one pull on one handle to lube the whole machine, instead of a dozen pumps of an alemite gun or oil can in a dozen different positions around the machine.

The oiling system inside a modern automotive or truck engine is similar in concept to the lube systems mentioned above, except that oil is pumped continuously. Much of this oil flows through passages drilled or cast into the engine block and cylinder heads, escaping through ports directly onto bearings, and squirting elsewhere to provide an oil bath. The oil pump simply pumps constantly, and any excess pumped oil continuously escapes through a relief valve back into the sump.

Many bearings in high-cycle industrial operations need periodic lubrication and cleaning, and many require occasional adjustment, such as pre-load adjustment, to minimise the effects of wear.

Bearing life is often much better when the bearing is kept clean and well lubricated. However, many applications make good maintenance difficult. For example, bearings in the conveyor of a rock crusher are exposed continually to hard abrasive particles. Cleaning is of little use, because cleaning is expensive yet the bearing is contaminated again as soon as the conveyor resumes operation. Thus, a good maintenance program might lubricate the bearings frequently but not include any disassembly for cleaning. The frequent lubrication, by its nature, provides a limited kind of cleaning action, by displacing older (grit-filled) oil or grease with a fresh charge, which itself collects grit before being displaced by the next cycle.

Rolling-Element Bearing Outer Race Fault Detection

Rolling-element bearings are widely used in the industries today, and hence maintenance of these bearings becomes an important task for the maintenance professionals. The rolling-element bearings wear out easily due to metal-to-metal contact, which creates faults in the outer race, inner race and ball. It is also the most vulnerable component of a machine because it is often under high load and high running speed conditions. Regular diagnostics of rolling-element bearing faults is critical for industrial safety and operations of the machines along with reducing the maintenance costs or avoiding shutdown time. Among the outer race, inner race and ball, the outer race tends to be more vulnerable to faults and defects.

There is still a room for discussion whether the rolling element excites the natural frequencies of bearing component when it passes the fault on the outer race. Hence we need to identify the bearing outer race natural frequency and its harmonics. The bearing faults create impulses and results in strong harmonics of the fault frequencies in the spectrum of vibration signals. These fault frequencies are sometimes masked by adjacent frequencies in the spectra due to their little energy. Hence, a very high spectral resolution is often needed to identify these frequencies during a FFT analysis. The natural frequencies of a rolling element bearing with the free boundary conditions are 3 kHz. Therefore, in order to use the bearing component resonance bandwidth method to detect the bearing fault at an initial stage a high frequency range accelerometer should be adopted, and data obtained from a long duration needs to be acquired. A fault characteristic frequency can only be identified when the fault extent is severe, such as that of a presence of a hole in the outer race. The harmonics of fault frequency is a more sensitive indicator of a bearing outer race fault. For a more serious detection of defected bearing faults waveform, spectrum and envelope techniques will help reveal these faults. However, if a high frequency demodulation is used in the envelope analysis in order to detect bearing fault characteristic frequencies, the maintenance professionals have to be more careful in the analysis because of resonance, as it may or may not contain fault frequency components.

Using spectral analysis as a tool to identify the faults in the bearings faces challenges due to issues like low energy, signal smearing, cyclostationarity etc. High resolution is often desired to differentiate the fault frequency components from the other high-amplitude adjacent frequencies. Hence, when the signal is sampled for FFT analysis, the sample length should be large enough to give adequate frequency resolution in the spectrum. Also, keeping the computation time and memory within limits and avoiding unwanted aliasing may be demanding. However, a minimal frequency resolution required can be obtained by estimating the bearing fault frequencies and other vibration frequency components and its harmonics due to shaft speed, misalignment, line frequency, gearbox etc.

Packing

Some bearings use a thick grease for lubrication, which is pushed into the gaps between the bearing surfaces, also known as *packing*. The grease is held in place by a plastic, leather, or rubber gasket (also called a *gland*) that covers the inside and outside edges of the bearing race to keep the grease from escaping.

Bearings may also be packed with other materials. Historically, the wheels on railroad cars used sleeve bearings packed with *waste* or loose scraps of cotton or wool fiber soaked in oil, then later used solid pads of cotton.

Ring Oiler

Bearings can be lubricated by a metal ring that rides loosely on the central rotating shaft of the bearing. The ring hangs down into a chamber containing lubricating oil. As the bearing rotates, viscous adhesion draws oil up the ring and onto the shaft, where the oil migrates into the bearing to lubricate it. Excess oil is flung off and collects in the pool again.

Splash Lubrication

Some machines contain a pool of lubricant in the bottom, with gears partially immersed in the liquid, or crank rods that can swing down into the pool as the device operates. The spinning wheels fling oil into the air around them, while the crank rods slap at the surface of the oil, splashing it randomly on the interior surfaces of the engine. Some small internal combustion engines specifically contain special plastic *flinger wheels* which randomly scatter oil around the interior of the mechanism.

Pressure Lubrication

For high speed and high power machines, a loss of lubricant can result in rapid bearing heating and damage due to friction. Also in dirty environments the oil can become contaminated with dust or debris that increases friction. In these applications, a fresh supply of lubricant can be continuously supplied to the bearing and all other contact surfaces, and the excess can be collected for filtration, cooling, and possibly reuse. Pressure oiling is commonly used in large and complex internal combustion engines in parts of the engine where directly splashed oil cannot reach, such as up into overhead valve assemblies. High speed turbochargers also typically require a pressurized oil system to cool the bearings and keep them from burning up due to the heat from the turbine.

Composite Bearings

Composite bearings are designed with a self-lubricating polytetrafluroethylene (PTFE) liner with a laminated metal backing. The PTFE liner offers consistent, controlled friction as well as durability whilst the metal backing ensures the composite bearing is robust and capable of withstanding high loads and stresses throughout its long life. Its design also makes it lightweight-one tenth the weight of a traditional rolling element bearing.

Types

There are many different types of bearings. Newer versions of more enabling designs are in development being tested, in which will reduce friction, increase bearing load, increase momentum build-up, and speed.

Type	Description	Friction	Stiffness†	Speed	Life	Notes
Plain bearing	Rubbing surfaces, usually with lubricant; some bearings use pumped lubrication and behave similarly to fluid bearings.	Depends on materials and construction, PTFE has coefficient of friction ~0.05-0.35, depending upon fillers added	Good, provided wear is low, but some slack is normally present	Low to very high	Low to very high - depends upon application and lubrication	Widely used, relatively high friction, suffers from stiction in some applications. Depending upon the application, lifetime can be higher or lower than rolling element bearings.
Rolling element bearing	Ball or rollers are used to prevent or minimise rubbing	Rolling coefficient of friction with steel can be ~0.005 (adding resistance due to seals, packed grease, preload and misalignment can increase friction to as much as 0.125)	Good, but some slack is usually present	Moderate to high (often requires cooling)	Moderate to high (depends on lubrication, often requires maintenance)	Used for higher moment loads than plain bearings with lower friction
Jewel bearing	Off-center bearing rolls in seating	Low	Low due to flexing	Low	Adequate (requires maintenance)	Mainly used in low-load, high precision work such as clocks. Jewel bearings may be very small.
Fluid bearing	Fluid is forced between two faces and held in by edge seal	Zero friction at zero speed, low	Very high	Very high (usually limited to a few hundred feet per second at/by seal)	Virtually infinite in some applications, may wear at startup/shutdown in some cases. Often negligible maintenance.	Can fail quickly due to grit or dust or other contaminants. Maintenance free in continuous use. Can handle very large loads with low friction.

Magnetic bearings	Faces of bearing are kept separate by magnets (electromagnets or eddy currents)	Zero friction at zero speed, but constant power for levitation, eddy currents are often induced when movement occurs, but may be negligible if magnetic field is quasi-static	Low	No practical limit	Indefinite. Maintenance free. (with electro-magnets)	Active magnetic bearings (AMB) need considerable power. Electrodynamic bearings (EDB) do not require external power.
Flexure bearing	Material flexes to give and constrain movement	Very low	Low	Very high.	Very high or low depending on materials and strain in application. Usually maintenance free.	Limited range of movement, no backlash, extremely smooth motion
Composite bearing	Plain bearing shape with PTFE liner on the interface between bearing and shaft with a laminated metal backing. PTFE acts as a lubricant.	PTFE and use of filters to dial in friction as necessary for friction control.	Good depending on laminated metal backing	Low to very high	Very high; PTFE and fillers ensure wear and corrosion resistance	Widely used, controls friction, reduces stick slip, PTFE reduces static friction
†Stiffness is the amount that the gap varies when the load on the bearing changes, it is distinct from the friction of the bearing.						

Cam Follower

In automotive terms a cam follower *may also refer to a tappet (or lifter) or rocker arm.*

A cam follower, also known as a track follower, is a specialized type of roller or needle bearing designed to follow cam lobe profiles. Cam followers come in a vast array of different configurations, however the most defining characteristic is how the cam follower mounts to its mating part; *stud* style cam followers use a stud while the *yoke* style has a hole through the middle.

Construction

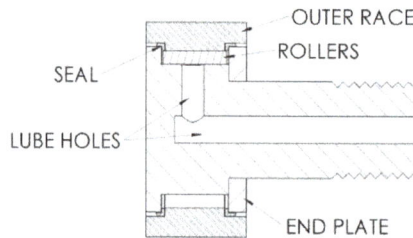

A cross-sectional view of a stud type cam follower

The modern stud type follower was invented and patented in 1937 by Thomas L. Robinson of the McGill Manufacturing Company. It replaced using a standard bearing and bolt. The new cam followers were easier to use because the stud was already included and they could also handle higher loads.

While roller cam followers are similar to roller bearings, there are quite a few differences. Standard ball and roller bearings are designed to be pressed into a rigid housing, which provides circumferential support. This keeps the outer race from deforming, so the race cross-section is relatively thin. In the case of cam followers the outer race is loaded at a single point, so the outer race needs a thicker cross-section to reduce deformation. However, in order to facilitate this the roller diameter must be decreased, which also decreases the dynamic bearing capacity.

End plates are used to contain the needles or bearing axially. On stud style followers one of the end plates is integrated into the inner race/stud; the other is pressed onto the stud up to a shoulder on the inner race. The inner race is induction hardened so that the stud remains soft if modifications need to be made. On yoke style followers the end plates are peened or pressed onto the inner race or liquid metal injected onto the inner race. The inner race is either induction hardened or through hardened.

Another difference is that a lubrication hole is provided to relubricate the follower periodically. A hole is provided at both ends of the stud for lubrication. They also usually have a black oxide finish to help reduce corrosion.

Types

There are many different types of cam followers available.

Anti-Friction Element

The most common anti-friction element employed is a *full complement* of needle rollers. This design can withstand high radial loads but no thrust loads. A similar design is the *caged needle roller* design, which also uses needle rollers, but uses a cage to keep them separated. This design allows for higher speeds but decreases the load capacity. The cage also increases internal space so it can hold more lubrication, which increases the time between relubrications. Depending on the exact design sometimes two rollers are put in each pocket of the cage, using a cage design originated by RBC Bearings in 1971.

For heavy-duty applications a *roller* design can be used. This employs two rows of rollers of larger diameter than used in needle roller cam followers to increase the dynamic load capacity and provide some thrust capabilities. This design can support higher speeds than the full complement design.

For light-duty applications a *bushing* type follower can be used. Instead of using a type of a roller a plastic bushing is used to reduce friction, which provides a maintenance free follower. The disadvantage is that it can only support light loads, slow speeds, no thrust loads, and the temperature limit is 200 °F (93 °C). A bushing type stud follower can only support approximately 25% of the load of a roller type stud follower, while the heavy and yoke followers can handle 50%. Also all-metallic heavy-duty bushing type followers exist.

Shape

The outer diameter (OD) of the cam follower (stud or yoke) can be the standard cylindrical shape or be crowned. Crowned cam followers are used to keep the load evenly distributed if it deflects or if there is any misalignment between the follower and the followed surface. They are also used in turntable type applications to reduce skidding. Crowned followers can compensate for up to 0.5° of misalignment, while a cylindrical OD can only tolerate 0.06°. The only disadvantage is that they cannot bear as much load because of higher stresses.

Stud

Stud style cam followers usually have a *standard* sized stud, but a *heavy* stud is available for increased static load capacity.

Drives

The standard driving system for a stud type cam follower is a slot, for use with a flat head screwdriver. However, hex sockets are available for higher torquing ability, which is especially useful for eccentric cam followers and those used in blind holes. Hex socket cam followers from most manufacturers eliminate the relubrication capability on that end of the cam follower. RBC Bearings' Hexlube cam followers feature a relubrication fitting at the bottom of the hex socket.

Eccentricity

Stud type cam followers are available with an eccentric stud. The stud has a bushing pushed onto it that has an eccentric outer diameter. This allows for adjustability during installation to eliminate any backlash. The adjustable range for an eccentric bearing is twice that of the eccentricity.

Yoke

Yoke type cam followers are usually used in applications where minimal deflection is required, as they can be support on both sides. They can support the same static load as a heavy stud follower.

Track Followers

All cam followers can be track followers, but not all track followers are cam followers. Some track followers have specially shaped outer diameters (OD) to follow tracks. For example, track followers are available with a V-groove for following a V-track, or the OD can have a flange to follow the lip of the track.

Specialized track followers are also designed to withstand thrust loads so the anti-friction elements are usually bearing balls or of a tapered roller bearing construction.

Mechanical Advantage

Mechanical advantage is a measure of the force amplification achieved by using a tool, mechanical device or machine system. Ideally, the device preserves the input power and simply trades off forces against movement to obtain a desired amplification in the output force. The model for this is the *law of the lever*. Machine components designed to manage forces and movement in this way are called mechanisms. An ideal mechanism transmits power without adding to or subtracting from it. This means the ideal mechanism does not include a power source, is frictionless, and is constructed from rigid bodies that do not deflect or wear. The performance of a real system relative to this ideal is expressed in terms of efficiency factors that take into account friction, deformation and wear.

The Law of The Lever

The lever is a movable bar that pivots on a fulcrum attached to or positioned on or across a fixed point. The lever operates by applying forces at different distances from the fulcrum, or pivot.

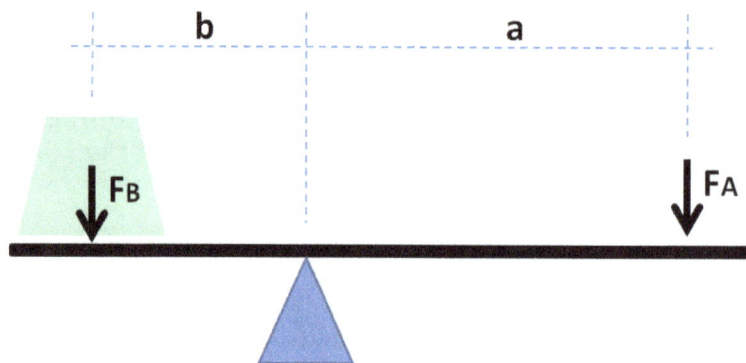

As the lever pivots on the fulcrum, points farther from this pivot move faster than points closer to the pivot. The power into and out of the lever must be the same. Power is the product of force and velocity, so forces applied to points farther from the pivot must be less than when applied to points closer in.

If a and b are distances from the fulcrum to points A and B and if force F_A applied to A is the input force and F_B exerted at B is the output, the ratio of the velocities of points A and B is given by a/b, so the ratio of the output force to the input force, or mechanical advantage, is given by

$$MA = \frac{F_B}{F_A} = \frac{a}{b}.$$

This is the *law of the lever*, which was proven by Archimedes using geometric reasoning. It shows that if the distance a from the fulcrum to where the input force is applied (point A) is greater than the distance b from fulcrum to where the output force is applied (point B), then the lever amplifies the input force. If the distance from the fulcrum to the input force is less than from the fulcrum to the output force, then the lever reduces the input force. Recognizing the profound implications and practicalities of the law of the lever, Archimedes has been famously attributed with the quotation "Give me a place to stand and with a lever I will move the whole world."

The use of velocity in the static analysis of a lever is an application of the principle of virtual work.

Speed Ratio

The requirement for power input to an ideal mechanism to equal power output provides a simple way to compute mechanical advantage from the input-output speed ratio of the system.

The power input to a gear train with a torque T_A applied to the drive pulley which rotates at an angular velocity of ω_A is $P = T_A \omega_A$.

Because the power flow is constant, the torque T_B and angular velocity ω_B of the output gear must satisfy the relation

$$P = T_A \omega_A = T_B \omega_B,$$

which yields

$$MA = \frac{T_B}{T_A} = \frac{\omega_A}{\omega_B}.$$

This shows that for an ideal mechanism the input-output speed ratio equals the mechanical advantage of the system. This applies to all mechanical systems ranging from robots to linkages.

Gear Trains

Gear teeth are designed so that the number of teeth on a gear is proportional to the radius of its pitch circle, and so that the pitch circles of meshing gears roll on each other without slipping. The speed ratio for a pair of meshing gears can be computed from ratio of the radii of the pitch circles and the ratio of the number of teeth on each gear, its gear ratio.

Two meshing gears transmit rotational motion.

The velocity v of the point of contact on the pitch circles is the same on both gears, and is given by

$$v = r_A \omega_A = r_B \omega_B,$$

where input gear A has radius r_A and meshes with output gear B of radius r_B, therefore,

$$\frac{\omega_A}{\omega_B} = \frac{r_B}{r_A} = \frac{N_B}{N_A}.$$

where N_A is the number of teeth on the input gear and N_B is the number of teeth on the output gear.

The mechanical advantage of a pair of meshing gears for which the input gear has N_A teeth and the output gear has N_B teeth is given by

$$MA = \frac{r_B}{r_A} = \frac{N_B}{N_A}.$$

This shows that if the output gear G_B has more teeth than the input gear G_A, then the gear train *amplifies* the input torque. And, if the output gear has fewer teeth than the input gear, then the gear train *reduces* the input torque.

If the output gear of a gear train rotates more slowly than the input gear, then the gear train is called a *speed reducer*. In this case, because the output gear must have more teeth than the input gear, the speed reducer will amplify the input torque.

Chain and Belt Drives

Mechanisms consisting of two sprockets connected by a chain, or two pulleys connected by a belt are designed to provide a specific mechanical advantage in power transmission systems.

The velocity v of the chain or belt is the same when in contact with the two sprockets or pulleys:

$$v = r_A \omega_A = r_B \omega_B,$$

where the input sprocket or pulley A meshes with the chain or belt along the pitch radius r_A and the output sprocket or pulley B meshes with this chain or belt along the pitch radius r_B,

therefore

$$\frac{\omega_A}{\omega_B} = \frac{r_B}{r_A} = \frac{N_B}{N_A}.$$

where N_A is the number of teeth on the input sprocket and N_B is the number of teeth on the output sprocket. For a toothed belt drive, the number of teeth on the sprocket can be used. For friction belt drives the pitch radius of the input and output pulleys must be used.

The mechanical advantage of a pair of a chain drive or toothed belt drive with an input sprocket with N_A teeth and the output sprocket has N_B teeth is given by

$$MA = \frac{T_B}{T_A} = \frac{N_B}{N_A}.$$

The mechanical advantage for friction belt drives is given by

$$MA = \frac{T_B}{T_A} = \frac{r_B}{r_A}.$$

Chains and belts dissipate power through friction, stretch and wear, which means the power output is actually less than the power input, which means the mechanical advantage of the real system will be less than that calculated for an ideal mechanism. A chain or belt drive can lose as much as 5% of the power through the system in friction heat, deformation and wear, in which case the efficiency of the drive is 95%.

Example: Bicycle Chain Drive

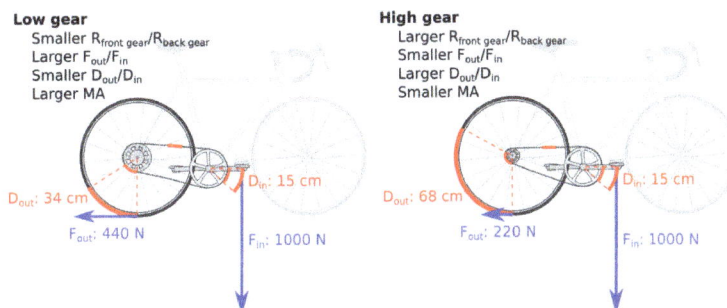

Mechanical advantage in different gears of a bicycle. Typical forces applied to the bicycle pedal and to the ground are shown, as are corresponding distances moved by the pedal and rotated by the wheel. Note that even in low gear the MA of a bicycle is less than 1.

Consider the 18-speed bicycle with 7 in (radius) cranks and 26 in (diameter) wheels. If the sprockets at the crank and at the rear drive wheel are the same size, then the ratio of the output force on the tire to the input force on the pedal can be calculated from the law of the lever to be

$$MA = \frac{F_B}{F_A} = \frac{7}{13} = 0.54.$$

Now, consider the small and large front sprockets which have 28 and 52 teeth respectively, and consider the small and large rear sprockets which have 16 and 32 teeth each. Using these numbers we can compute the following speed ratios between the front and rear sprockets

Speed ratios							
	input (small)	input (large)	output (small)	output (large)	speed ratio	crank-wheel ratio	total MA
low speed	28	45	19	32	1.14	0.54	0.62
mid 1	19	52	29	32	0.62	0.54	0.33
mid 2	28	39	16	26	0.57	0.54	0.31
high speed	-	52	16	-	0.30	0.54	0.16

The ratio of the force driving the bicycle to the force on the pedal, which is the total mechanical advantage of the bicycle, is the product of the speed ratio and the crank-wheel lever ratio.

Notice that in every case the force on the pedals is greater than the force driving the bicycle forward (in the illustration above, the corresponding backward-directed reaction force on the ground is indicated). This low mechanical advantage keeps the pedal crank speed low relative to the speed of the drive wheel, even in low gears.

Block and Tackle

A block and tackle is an assembly of a rope and pulleys that is used to lift loads. A number of pulleys are assembled together to form the blocks, one that is fixed and one that moves with the load. The rope is threaded through the pulleys to provide mechanical advantage that amplifies that force applied to the rope.

In order to determine the mechanical advantage of a block and tackle system consider the simple case of a gun tackle, which has a single mounted, or fixed, pulley and a single movable pulley. The rope is threaded around the fixed block and falls down to the moving block where it is threaded around the pulley and brought back up to be knotted to the fixed block.

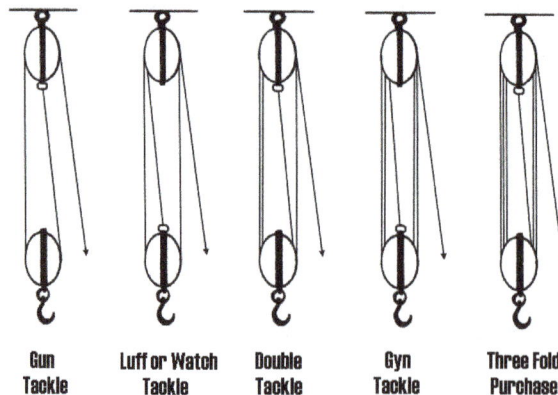

Gun Tackle   Luff or Watch Tackle   Double Tackle   Gyn Tackle   Three Fold Purchase

The mechanical advantage of a block and tackle equals the number of sections of rope that support the moving block; shown here it is 2, 3, 4, 5, and 6, respectively.

Let S be the distance from the axle of the fixed block to the end of the rope, which is A where the input force is applied. Let R be the distance from the axle of the fixed block to the axle of the moving block, which is B where the load is applied.

The total length of the rope L can be written as

$$L = 2R + S + K,$$

where K is the constant length of rope that passes over the pulleys and does not change as the block and tackle moves.

The velocities V_A and V_B of the points A and B are related by the constant length of the rope, that is

$$\dot{L} = 2\dot{R} + \dot{S} = 0,$$

or

$$\dot{S} = -2\dot{R}.$$

The negative sign shows that the velocity of the load is opposite to the velocity of the applied force, which means as we pull down on the rope the load moves up.

Let V_A be positive downwards and V_B be positive upwards, so this relationship can be written as the speed ratio

$$\frac{V_A}{V_B} = \frac{\dot{S}}{-\dot{R}} = 2,$$

where 2 is the number of rope sections supporting the moving block.

Let F_A be the input force applied at A the end of the rope, and let F_B be the force at B on the moving block. Like the velocities F_A is directed downwards and F_B is directed upwards.

For an ideal block and tackle system there is no friction in the pulleys and no deflection or wear in the rope, which means the power input by the applied force $F_A V_A$ must equal the power out acting on the load $F_B V_B$, that is

$$F_A V_A = F_B V_B.$$

The ratio of the output force to the input force is the mechanical advantage of an ideal gun tackle system,

$$MA = \frac{F_B}{F_A} = \frac{V_A}{V_B} = 2.$$

This analysis generalizes to an ideal block and tackle with a moving block supported by n rope sections,

$$MA = \frac{F_B}{F_A} = \frac{V_A}{V_B} = n.$$

This shows that the force exerted by an ideal block and tackle is n times the input force, where n is the number of sections of rope that support the moving block.

Efficiency

Mechanical advantage that is computed using the assumption that no power is lost through deflection, friction and wear of a machine is the maximum performance that can be achieved. For this reason, it is often called the *ideal mechanical advantage* (IMA). In operation, deflection, friction and wear will reduce the mechanical advantage. The amount of this reduction from the ideal to the *actual mechanical advantage* (AMA) is defined by a factor called *efficiency*, a quantity which is determined by experimentation.

As an ideal example, using a block and tackle with six ropes and a 600 pound load, the operator would be required to pull the rope six feet and exert 100 pounds of force to lift the load one foot. Both the ratios F_{out} / F_{in} and V_{in} / V_{out} from below show that the IMA is six. For the first ratio, 100 pounds of force in results in 600 pounds of force out; in the real world, the force out would be less than 600 pounds. The second ratio also yields a MA of 6 in the ideal case but fails in real world calculations; it does not properly account for energy losses. Subtracting those losses from the IMA or using the first ratio yields the AMA. The ratio of AMA to IMA is the mechanical efficiency of the system.

Ideal Mechanical Advantage

The *ideal mechanical advantage* (IMA), or *theoretical mechanical advantage*, is the mechanical advantage of a device with the assumption that its components do not flex, there is no friction, and there is no wear. It is calculated using the physical dimensions of the device and defines the maximum performance the device can achieve.

The assumptions of an ideal machine are equivalent to the requirement that the machine does not store or dissipate energy; the power into the machine thus equals the power out. Therefore, the power P is constant through the machine and force times velocity into the machine equals the force times velocity out--that is,

$$P = F_{in} v_{in} = F_{out} v_{out}.$$

The ideal mechanical advantage is the ratio of the force out of the machine (load) to the force into the machine (effort), or

$$IMA = \frac{F_{out}}{F_{in}}.$$

Applying the constant power relationship yields a formula for this ideal mechanical advantage in terms of the speed ratio:

$$IMA = \frac{F_{out}}{F_{in}} = \frac{v_{in}}{v_{out}}.$$

The speed ratio of a machine can be calculated from its physical dimensions. The assumption of constant power thus allows use of the speed ratio to determine the maximum value for the mechanical advantage.

Actual Mechanical Advantage

The *actual mechanical advantage* (AMA) is the mechanical advantage determined by physical measurement of the input and output forces. Actual mechanical advantage takes into account energy loss due to deflection, friction, and wear.

The AMA of a machine is calculated as the ratio of the measured force output to the measured force input,

$$AMA = \frac{F_{out}}{F_{in}},$$

where the input and output forces are determined experimentally.

The ratio of the experimentally determined mechanical advantage to the ideal mechanical advantage is the efficiency η of the machine,

$$\eta = \frac{AMA}{IMA}.$$

Mechanical Advantage Device

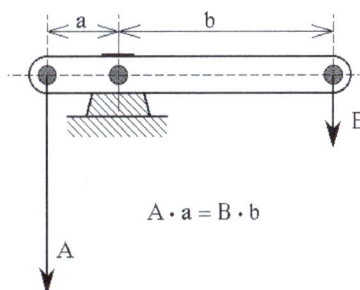

$$A \cdot a = B \cdot b$$

Beam balanced around a fulcrum

A simple machine that exhibits mechanical advantage is called a mechanical advantage device - e.g.:

- Lever: The beam shown is in static equilibrium around the fulcrum. This is due to the moment created by vector force "A" counterclockwise (moment A*a) being in equilibrium with the moment created by vector force "B" clockwise (moment B*b). The relatively low vector force "B" is translated in a relatively high vector force "A". The force is thus increased in the ratio of the forces A : B, which is equal to the ratio of the distances to the fulcrum b : a. This ratio is called the mechanical advantage. This idealised situation does not take into account friction.

- Wheel and axle motion (e.g. screwdrivers, doorknobs): A wheel is essentially a lever with one arm the distance between the axle and the outer point of the wheel, and the other the radius of the axle. Typically this is a fairly large difference, leading to a proportionately large mechanical advantage. This allows even simple wheels with wooden axles running in wooden blocks to still turn freely, because their friction is overwhelmed by the rotational force of the wheel multiplied by the mechanical advantage.

- A block and tackle of multiple pulleys creates mechanical advantage, by having the flexible material looped over several pulleys in turn. Adding more loops and pulleys increases the mechanical advantage.

- Screw: A screw is essentially an inclined plane wrapped around a cylinder. The run over the rise of this inclined plane is the mechanical advantage of a screw.

Pulleys

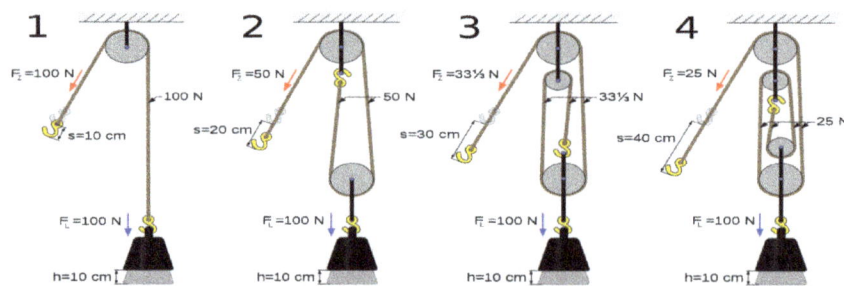

Examples of rope and pulley systems illustrating mechanical advantage.

Consider lifting a weight with rope and pulleys. A rope looped through a pulley attached to a fixed spot, e.g. a barn roof rafter, and attached to the weight is called a *single pulley*. It has an mechanical advantage (MA) = 1 (assuming frictionless bearings in the pulley), moving no mechanical advantage (or disadvantage) however advantageous the change in direction may be.

A *single movable pulley* has an MA of 2 (assuming frictionless bearings in the pulley). Consider a pulley attached to a weight being lifted. A rope passes around it, with one end attached to a fixed point above, e.g. a barn roof rafter, and a pulling force is applied upward to the other end with the two lengths parallel. In this situation the distance the lifter must pull the rope becomes twice the distance the weight travels, allowing the force applied to be halved. Note: if an additional pulley is used to change the direction of the rope, e.g. the person doing the work wants to stand on the ground instead of on a rafter, the mechanical advantage is not increased.

By looping more ropes around more pulleys we can continue to increase the mechanical advantage. For example, if we have two pulleys attached to the rafter, two pulleys attached to the weight, one end attached to the rafter, and someone standing on the rafter pulling the rope, we have a mechanical advantage of four. Again note: if we add another pulley so that someone may stand on the ground and pull down, we still have a mechanical advantage of four.

Here are examples where the fixed point is not obvious:

- A velcro strap on a shoe passes through a slot and folds over on itself. The slot is a movable pulley and the MA = 2.

- Two ropes laid down a ramp attached to a raised platform. A barrel is rolled onto the ropes and the ropes are passed over the barrel and handed to two workers at the top of the ramp. The workers pull the ropes together to get the barrel to the top. The barrel is a movable pulley and the MA = 2. If there is enough friction where the rope is pinched between the barrel and the ramp, the pinch point becomes the attachment point. This is considered a fixed attachment point because the rope above the barrel does not move relative to the ramp. Alternatively the ends of the rope can be attached to the platform.

- Block and tackle: MA = 2 or more, depending on design.

Screws

The theoretical mechanical advantage for a screw can be calculated using the following equation:

$$MA = \frac{\pi d_m}{l}$$

where

d_m = the mean diameter of the screw thread

l = the lead of the screw thread

- Note that the actual mechanical advantage of a screw *system* is greater, as a screwdriver or other screw driving system has a mechanical advantage as well.

Inclined plane: MA = length of slope ÷ height of the slope

Using a Rope as a Lever

You understand what a lever is. But can you use a rope as a lever? You can if you are using is as shown in the figure. Arrange a strong column at a steep angle, apply a firm attachment from the rope to the column and the object to be lifted, and apply the force horizontally, and get a great multiplier of force. If one combines the mechanical advantage of pulleys in the horizontal force applier, you could get even greater lifting force. Picture the application in the figure, and in this case, you could get great lift upward then add planks of wood to support the load at height, reset, and lift again, and add more blocking, and lift, repeat. This technology has been lost in history, as the use of winches with low friction wheeled pulleys and reduction gearing powered by steam and then electricity (along with steel cable replacing rope) took over this primitive technology.

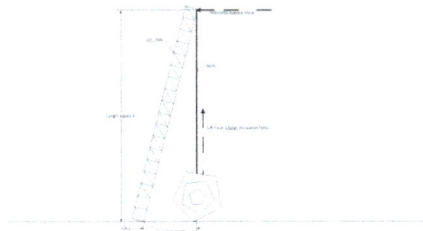

A sample diagram of using a rope to lift a heavy object with the mechanical advantage of geometry, i.e. a lever.

References

- Fisher, Len (2003), How to Dunk a Doughnut: The Science of Everyday Life, Arcade Publishing, ISBN 978-1-55970-680-3.

- United States Bureau of Naval Personnel (1971), Basic machines and how they work (Revised 1994 ed.), Courier Dover Publications, ISBN 978-0-486-21709-3.

- Usher, A. P. (1929). A History of Mechanical Inventions. Harvard University Press (reprinted by Dover Publications 1988). p. 94. ISBN 978-0-486-14359-0. OCLC 514178. Retrieved 7 April 2013.

- Paul, Akshoy; Roy, Pijush; Mukherjee, Sanchayan (2005), Mechanical sciences: engineering mechanics and strength of materials, Prentice Hall of India, p. 215, ISBN 81-203-2611-3.

- Asimov, Isaac (1988), Understanding Physics, New York, New York, USA: Barnes & Noble, p. 88, ISBN 0-88029-251-2.

- Usher, Abbott Payson (1988). A History of Mechanical Inventions. USA: Courier Dover Publications. p. 98. ISBN 0-486-25593-X.

- Chiu, Y. C. (2010), An introduction to the History of Project Management, Delft: Eburon Academic Publishers, p. 42, ISBN 90-5972-437-2

- Ostdiek, Vern; Bord, Donald (2005). Inquiry into Physics. Thompson Brooks/Cole. p. 123. ISBN 0-534-49168-5. Retrieved 2008-05-22.

- Krebs, Robert E. (2004). Groundbreaking Experiments, Inventions, and Discoveries of the Middle Ages. Greenwood Publishing Group. p. 163. ISBN 0-313-32433-6. Retrieved 2008-05-21.

- Stephen, Donald; Lowell Cardwell (2001). Wheels, clocks, and rockets: a history of technology. USA: W. W. Norton & Company. pp. 85–87. ISBN 0-393-32175-4.

- Bhatnagar, V. P. (1996). A Complete Course in Certificate Physics. India: Pitambar Publishing. pp. 28–30. ISBN 8120908686.

- Simmons, Ron; Cindy Barden (2008). Discover! Work & Machines. USA: Milliken Publishing. p. 29. ISBN 1429109475.

- Uicker, Jr., John J.; Pennock, Gordon R.; Shigley, Joseph E. (2003), Theory of Machines and Mechanisms (third ed.), New York: Oxford University Press, ISBN 978-0-19-515598-3

- Goyal, M. C.; G. S. Raghuvanshi (2009). Engineering Mechanics. New Delhi: PHI Learning Private Ltd. p. 202. ISBN 81-203-3789-1.

- Simionescu, P.A. (2014). Computer Aided Graphing and Simulation Tools for AutoCAD users (1st ed.). Boca Raton, FL: CRC Press. ISBN 9-781-48225290-3.

- Bhandari, V.B. (2010). Design of machine elements. Tata McGraw-Hill. p. 472. ISBN 9780070681798. Retrieved 9 February 2016.

- Guran, Ardéshir; Rand, Richard H. (1997), Nonlinear dynamics, World Scientific, p. 178, ISBN 978-981-02-2982-5.

- White, John H. (1985) [1978]. The American Railroad Passenger Car. 2. Baltimore, MD: Johns Hopkins University Press. p. 518. ISBN 0801827477. OCLC 11469984.

Permissions

All chapters in this book are published with permission under the Creative Commons Attribution Share Alike License or equivalent. Every chapter published in this book has been scrutinized by our experts. Their significance has been extensively debated. The topics covered herein carry significant information for a comprehensive understanding. They may even be implemented as practical applications or may be referred to as a beginning point for further studies.

We would like to thank the editorial team for lending their expertise to make the book truly unique. They have played a crucial role in the development of this book. Without their invaluable contributions this book wouldn't have been possible. They have made vital efforts to compile up to date information on the varied aspects of this subject to make this book a valuable addition to the collection of many professionals and students.

This book was conceptualized with the vision of imparting up-to-date and integrated information in this field. To ensure the same, a matchless editorial board was set up. Every individual on the board went through rigorous rounds of assessment to prove their worth. After which they invested a large part of their time researching and compiling the most relevant data for our readers.

The editorial board has been involved in producing this book since its inception. They have spent rigorous hours researching and exploring the diverse topics which have resulted in the successful publishing of this book. They have passed on their knowledge of decades through this book. To expedite this challenging task, the publisher supported the team at every step. A small team of assistant editors was also appointed to further simplify the editing procedure and attain best results for the readers.

Apart from the editorial board, the designing team has also invested a significant amount of their time in understanding the subject and creating the most relevant covers. They scrutinized every image to scout for the most suitable representation of the subject and create an appropriate cover for the book.

The publishing team has been an ardent support to the editorial, designing and production team. Their endless efforts to recruit the best for this project, has resulted in the accomplishment of this book. They are a veteran in the field of academics and their pool of knowledge is as vast as their experience in printing. Their expertise and guidance has proved useful at every step. Their uncompromising quality standards have made this book an exceptional effort. Their encouragement from time to time has been an inspiration for everyone.

The publisher and the editorial board hope that this book will prove to be a valuable piece of knowledge for students, practitioners and scholars across the globe.

Index

www.ingramcontent.com/pod-product-compliance
Lightning Source LLC
Chambersburg PA
CBHW061312190326
41458CB00011B/3785